51. K.V. Novozhilov (Chief editor): *Microbiological Methods for Biological Control of Pests of Agricultural Crops*
52. K.I. Rossinskii (editor): *Dynamics and Thermal Regimes of Rivers*
53. K.V. Gnedin: *Operating Conditions and Hydraulics of Horizontal Settling Tanks*
54. G.A. Zakladnoi & V.F. Ratanova: *Stored-grain Pests and Their Control*
55. Ts.E. Mirtskhulava: *Reliability of Hydro-reclamation Installations*
56. Ia. S. Ageikin: *Off-the-road Mobility of Automobiles*
57. A.A. Kmito & Yu.A. Sklyarov: *Pyrheliometry*
58. N.S. Motsonelidze: *Stability and Seismic Resistance of Buttress Dams*
59. Ia.S. Ageikin: *Off-the-road Wheeled and Combined Traction Devices*
60. Iu.N. Fadeev & K.V. Novozhilov: *Integrated Plant Protection*
61. N.A. Izyumova: *Parasitic Fauna of Reservoir Fishes of the USSR and Its Evolution*
62. O.A. Skarlato (Editor-in-Chief): *Investigation of Monogeneans in the USSR*
63. A.I. Ivanov: *Alfalfa*
64. Z.S. Bronshtein: *Fresh-water Ostracoda*
65. M.G. Chukhrii: *An Atlas of the Ultrastructure of Viruses of Lepidopteran Pests of Plants*
66. E.S. Bosoi et al.: *Theory, Construction and Calculations of Agricultural Machines,* Volume 1
67. G.A. Avsyuk (Editor-in-Chief): *Data of Glaciological Studies*
68. G.A. Mchedlidze: *Fossil Cetacea of the Caucasus*
69. A.M. Akramkhodzhaev: *Geology and Exploration of Oil- and Gas-bearing Ancient Deltas*
70. N.M. Berezina & D.A. Kaushanskii: *Presowing Irradiation of Plant Seeds*
71. G.U. Lindberg & Z.V. Krasyukova: *Fishes of the Sea of Japan and the Adjacent Deltas*
72. N.I. Plotnikov & I.I. Roginets: *Hydrogeology of Ore Deposits*
73. A.V. Balushkin: *Morphological Bases of the Systematics and Phylogeny of the Nototheniid Fishes*
74. E.Z. Pozin et al.: *Coal Cutting by Winning Machines*
75. S.S. Shul'man: *Myxosporidia of the USSR*
76. G.N. Gogonenkov: *Seismic Prospecting for Sedimentary Formations*
77. I.M. Batugina & I.M. Petukhov: *Geodynamic Zoning of Mineral Deposits for Planning and Exploitation of Mines*
78. I.I. Abramovich & I.G. Klushin: *Geodynamics and Metallogeny of Folded Belts*
79. M.V. Mina: *Microevolution of Fishes*
80. K.V. Konyaev: *Spectral Analysis of Physical Oceanographic Data*
81. A.I. Tseitlin & A.A. Kusainov: *Role of Internal Friction in Dynamic Analysis of Structures*
82. E.A. Kozlov: *Migration in Seismic Prospecting*
83. E.S. Bosoi et al.: *Theory, Construction and Calculations of Agricultural Machines,* Volume 2
84. B.B. Kudryashov and A.M. Yakovlev: *Drilling in the Permafrost*
85. T.T. Klubova: *Clayey Reservoirs of Oil and Gas*
86. G.I. Amurskii et al.: *Remote-sensing Methods in Studying Tectonic Fractures in Oil- and Gas-bearing Formations*
87. A.V. Razvalyaev: *Continental Rift Formation and Its Prehistory*
88. V.A. Ivovich & L.N. Pokrovskii: *Dynamic Analysis of Suspended Roof Systems*
89. N.P. Kozlov (Technical Editor): *Earth's Nature from Space*
90. M.M. Grachevskii & A.S. Kravchuk: *Hydrocarbon Potential of Oceanic Reefs of the World*
91. K.V. Mikhailov et al.: *Polymer Concretes and Their Structural Uses*

POLYMER CONCRETES
AND
THEIR STRUCTURAL USES

POLYMER CONCRETES
AND
THEIR STRUCTURAL USES

K.V. MIKHAILOV
V.V. PATUROEV
USSR

R. KREIS
Germany

Russian Translations Series
91

A.A. BALKEMA/ROTTERDAM/BROOKFIELD/1992

Translation of: *Polimerbetoni i konstruktsii na ikh osnove, Stroiizdat, Moscow, 1989*

© 1992 Copyright reserved

Translator : P.M. Rao
Technical Editor : V.S. Parameswaran
General Editor : Margaret Majithia

ISBN 90 6191 110 9

Preface

The advancement of the building industry is associated with continuous research, development, and design of new and improved materials and construction of structures utilising them and possessing a number of efficient features, such as good decorative qualities, high strength and chemical resistance to various aggressive media.

The 20th century can rightly be called not only the century of nuclear power, space, and electronics, but also the age of polymers. According to long-term scientific and technical forecasts, the world production of polymers would outstrip the consumption of ferrous metals by volume as of 1990, and by the end of the first decade of the 21st century, by weight also.

The building industry, the largest consumer of polymers, consumes 25 to 30% of the total volume of polymers produced.

One of the efficient fields in the application of polymers in the building industry is the development of innumerable types of polymer concretes by mixing a polymer binder and mineral aggregates followed by their hardening. Polymer binders represent compositions consisting of synthetic monomers or oligomers, hardeners, modifying additives, and finely dispersed fillers. Depending on the composition, polymer concretes may possess high strength, density and chemical resistance to most industrial aggressive media, and high decorative and finishing features, imitating marble, onyx, decorative granite, and other building stones. Electrically conducting polymer concretes and those with high dielectric and damping properties have been developed. Unlike industrial plastics in which the polymer content varies from 50 to 70%, polymer concretes contain only 5 to 10% polymers by weight, while the remainder is made up of mineral aggregates and fillers.

A very high economy has been realised by using polymer concretes in the building industry. Nevertheless, polymer concretes represent comparatively new constructional materials of complex composition which have not yet been investigated fully. The detailed and comprehensive investigations on polymer concretes carried out by the authors and an analysis of the numerous investigations carried out by several other researchers form the basis of this publication, in which the data available on these efficient materials have been drawn together.

Taking into consideration the characteristic features of design and technology of producing structures and products based on polymer concretes, this book has assigned considerable importance to the theory of structural formation and selection of optimal compositions of polymer concretes, physical and chemical characteristics that have a direct bearing on production technology, and the classification of degradative processes. Details of efficient equipment, techniques and case studies of production lines, and descriptions of a variety of interesting products and structures and the appropriate fields of their application have been included.

The authors hope that this book will contribute to the further advancement of industrial production of polymer concrete structures and their application in the areas of housing, civil, and industrial construction, and help the designers to choose and apply more efficiently the appropriate types of polymer concretes in their designs.

Chapters 1, 3, 5, 6, 7, and 10 have been written by V.V. Paturoev, Chapter 8 by K.V. Mikhailov, and Chapters 2, 4, 9, and 11 jointly by R. Kreis and V.V. Paturoev.

Contents

1

General Information and Classification of Polymer Concretes

1.1. General Information on Polymer Concretes

Polymer concretes represent a new generation of efficient and chemical-resistant materials in which the mineral fillers and aggregates reach 90 to 95% by weight. The content of the polymer binder in these comparatively new materials is only 5 to 10% of the total weight of the polymer concrete. The cost of polymer concretes is therefore far less than that of plastics.

In spite of such a comparatively low consumption of polymer binder per unit weight, polymer concretes possess high density, strength, chemical resistance, and many other desirable features. The appropriate selection of binder, fillers, and aggregates helps to produce polymer concretes with high dielectric characteristics or, on contrarily, those possessing good electrical conductivity, vacuum tightness, or damping characteristics. Compositions of special concretes with excellent properties of protection from various kinds of radiation have been developed. Further, the high proportion of fillers used in such concretes sharply reduces shrinkage (which is equal to that of cement concretes), while the modulus of elasticity is greatly enhanced. This permits the use of such concretes in load-bearing and other vital structural members and also in machine tool manufacturing and engineering industries. For example, compositions of polymer concretes having a density of 2200 to 2400 kg/m^3 possess the following compressive strengths: based on phenol-formaldehyde resins 40–60 MPa, carbamide resins 50–80 MPa, polyester and methyl methacrylate resins 80–120 MPa, epoxide resins up to 150 MPa, vinyl esters and furan-epoxide resins up to 190 MPa, and so forth.

Depending on the type of mineral fillers and aggregates, heat-insulating polymer concretes with a density ranging from 300–400 to 800–1000 kg/m^3, light polymer concretes with a density up to 1800–2000 kg/m^3 and superheavy polymer concretes with a density up to 5500 kg/m^3 can be produced [50, 51, 62].

The results of investigations show that the above physical and mechanical characteristics of polymer concretes are not the maximum possible values and technological improvements can be expected to surpass these values significantly in the very near future. The great advances made in the chemical sciences and by the industry in the field of synthesising new types of monomers and oligomers, many of which possess unique properties, point to such a possibility.

It is quite well known that one of the vital drawbacks of polymer concretes is their comparatively low thermal stability (80 to 120°C). Organosilicon and other binders which have already been developed can be used to produce polymer concretes with a thermal stability of up to 600°C or more.

Polymer concretes were first used mainly as decorative and finishing materials and chemically stable structural components and members. Subsequently, the areas of their application enlarged continuously to such an extent that they are now effectively used in extremely diverse fields, such as the building and electrical engineering industries, radio electronics and atomic power, land reclamation, machine tool manufacturing and engineering industries.

Some ten types of monomers or oligomers are used throughout the world in various combinations along with modifiers to yield over thirty varieties of polymer concretes. The more extensively used are those based on polyesters and epoxide resins, vinyl esters and methyl methacrylate monomers. Phenol resins are used less frequently. Apart from these resins, polymer concretes based on furan, furan-epoxide, carbamide, and phenol-formaldehyde resins are also quite extensively used in the Soviet Union.

An analysis of the literature and the reports of five international congresses on the use of polymers in concrete revealed a perceptibly increasing interest in these materials not only among specialists and scientists, but also a wide group of representatives from the industry. This is because the experience gained in using the various types of polymer concretes in extremely diverse fields has proved their positive features and cost effectiveness.

The production and extensive use in many countries of decorative and finishing wall and window panels, staircases and other products using polymer concretes imitative of marble, onyx, and other high-quality natural building stones, suffices as proof of their cost effectiveness. Mention should also be made of the use of polymer concretes in the manufacturing of pipes, silos, chemically resistant building appurtenances, high-voltage insulators, containers for storing aggressive liquids, baths for electrolysis of non-ferrous metals, underwater submersibles, etc.

A new and highly promising field is the use of polymer concretes in the making of body components of reducers, centrifugal pumps, and other apparatus in the engineering industry, frames of high-precision lathes, and pattern plates for machine tool manufacture.

In damping properties, polymer concretes are 5 to 6 times superior to cast iron and withstand the action of oils and cooling liquids well; they require no additional painting[1].

Chemically stable polymer concretes based on furan resins and reinforced with steel were produced in the Soviet Union for the first time for use in load-bearing structures (columns of platforms supporting baths, foundation blocks, beams, crossbars, etc.).

An inspection of polymer concrete columns of platforms supporting baths in a copper electrolysis shop (1986) revealed no sign of damage after 17 years of use. Reinforced concrete columns under such conditions have to be given an annual protective coating with a paste based on epoxide resin.

In spite of a much higher initial cost, there is an actual economy of 400 to 500 rubles per m³ when using polymer concrete structures.

Calculations of western specialists showed that the specific energy consumption for producing a unit weight of polymer concrete is 2.5, steel 5–7, porcelain for insulators 5–10, and aluminium 7.5–10, assuming it as 1 for cement concrete. Further, by introducing the coefficient of economic efficiency, which represents the ratio of economy as a result of the improved properties to the cost of the material, and assuming it as equal to one for cement concrete, it goes up to four or more for polymer concretes. This coefficient will go up to 10 to 12 for columns of platforms supporting tanks and electrolysis baths of non-ferrous metals and other structures.

An extremely interesting form of light polymer concrete has been developed by Oder Ferwaltung Company (Federal Republic of Germany) under the name 'Gralitbeton'. This is a fine-grained polymer concrete based on phenol-formaldehyde resin:

Density, kg/m³	1600–1700
Ultimate strength, MPa:	
compressive	35
bending	11–12
Water absorption after 24 hr, %	16–18

According to this company, these characteristics of 'Gralitbeton' were realised by a very small consumption of comparatively cheap phenol resins at 3.5–4% of the total weight of polymer concretes, equivalent to 60–70 kg of resin per m³ of concrete.

This company developed the technology and the necessary equipment for production. The range of products and structural members produced includes panels for internal partitions, building blocks, facing panels, etc. This polymer concrete possesses the following advantages compared to other types of polymer concretes: small number of constituents (absence of coarse fractions) because it

---

[1] More detailed information on this subject is given in Chapter 9.

is essentially a fine-grained sandy polymer concrete and high strength characteristics while consuming a very small quantity of polymer binder at low density.

The main drawback of this concrete is its high water absorption and hence low cold resistance. Hence the company mainly produces products and structures for countries in the Near East (Saudi Arabia, Kuwait, Israel, etc.).

A polymer concrete based on a complex binder with a thermal stability of up to 250 to 300°C has been produced at the Brookhaven National Laboratory (USA). This polymer concrete is intended for lining metal pipes used in drilling geothermal wells.

In the Federal Republic of Germany, Soviet Union, and many other countries, apart from polymer concretes designed for producing chemically resistant load-bearing structures and decorative-finishing articles, several compositions have been developed for machine tool manufacture and engineering industries using polymer concretes that possess high dielectric and electrical conducting properties, and vacuum-tight and heat-resistant compositions that are resistant to various types of radiations.

More detailed information on these polymer concretes has been given in the appropriate sections of this monograph.

1.2. Classification of Polymer Concretes

The search for ways and means of increasing the strength, density, chemical resistance, and durability of plain and reinforced concrete has led to the formulation of a wide group of concretes with additives or based on polymers, called polymer concretes [50, 51, 62].

The use of polymers in concrete involves three fundamental principles: full or partial replacement of inorganic by organic aggregates; total or partial replacement of inorganic by organic binder; and introduction of organic binder into the porous structure of the cement, stone or concretes.

The substitution of inorganic by organic filler is directed towards obtaining light concretes, reducing their average density, and improving their heat-insulating characteristics. The properties of concrete using organic aggregates, for example, foam polystyrene, are practically no different from those which could be produced with inorganic aggregates, such as foam glass, claydite (clay filler), azurite, etc.

A very complex picture emerges on the interaction of synthetic resins and cement used as binder, especially on totally replacing the inorganic by organic binder. In this case, material of a new type with special properties is formed.

Cement-free concretes using polymer binders (polymer concretes) were initially given several names. For example, in the Soviet Union, they were called 'plastic concretes', 'organomineral concretes', etc. In the western countries, practically every company producing polymer concrete components gave trade names to them. Thus, in the Federal Republic of Germany, polymer concretes were called 'Duroplast', 'Degadur', 'Dezament', 'Plexilite', etc. Such arbitrary terminology contributed to some confusion, sometimes rendering it difficult to comprehend the phenomena described.

We made the first attempt in 1968 to develop a unified classification and terminology applicable to polymer concretes and later improved upon it. This classification has presently been recognised in the Soviet Union as well as abroad and has been utilised in compiling Russian–French, Russian–English, and Russian–English–Japanese dictionaries of technical terms [41, 46, 62, 68].

It should be pointed out that, apart from a general classification, polymer concretes have also been classified according to the ease of application of the mixes, types of polymer binders and hardeners and degradation processes in polymer concretes under the influence of various aggressive media and physical effects.

According to the general classification developed by the authors, special concretes with additives or those based on polymers are divided into four main categories depending on the composition and the method of production: concretes based on polymer binders—polymer concretes (PC); cement concretes modified by polymers—polymer-cement concretes (PCC); sulphur concretes modified by polymers—polymer sulphur concretes (PSC); and cement concretes impregnated with monomers or oligomers—concrete polymers (CP) (Fig. 1.1).

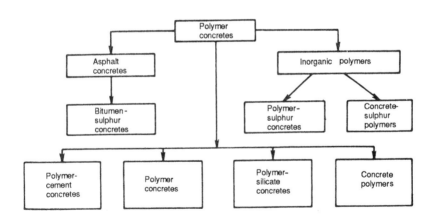

Fig. 1.1. Classification of polymer concretes.

Polymer concretes denote compositions based on synthetic resins or monomers and chemically stable fillers and aggregates without using mineral binders and water. The composition of polymer concretes consists of no less than three fractions of fillers and aggregates: finely dispersed fillers with a particle size of less than 0.15 mm; aggregates—sand with a grain size up to 5 mm; and rubble up to 50 mm in size. Unlike polymer concretes, polymer slurries do not contain aggregates. Pastes contain only one dispersed filler fraction.

As will be explained later, the main properties of polymer concretes are determined by the chemical nature of the synthetic resin and the type and content of the finely dispersed filler fraction. Coarse fractions of aggregates (sand and stone aggregates), while essentially performing the role of a skeleton, influence the main physical and mechanical properties to a small extent. Therefore, in the case of non-reinforced material, the abbreviation of the name of the polymer binder and the type of finely dispersed filler are shown after the word 'polymer concrete'; for reinforced material, the type of reinforcement is shown before the name of the material, e.g., polymer concrete FAM (grade of furan resin) based on andesite, polymer concrete PN (polyester maleinate resin) using marshalite (silica flour), steel polymer concrete FAM using agloporite, etc.

Polymer cement concretes. These are cement concretes during the production of which organosilicon or water-soluble polymers, aqueous emulsions of the type of polyvinyl acetate, or water-soluble epoxide resins are added to the mix at 2–3 to 18–20%.

Polymer sulphur concretes. These are concretes based on a sulphur binder during the production of which modifying additives of the type of dicyclopentadiene or chloroparaffin are added to the molten sulphur at 1–2 to 12–15%.

Concrete polymers. These are cement concretes which, after the completion of hardening and structure formation processes, are subjected to drying and impregnated with various monomers or oligomers in the porous structure of the concrete followed by radiation or thermocatalytic polymerization. The impregnation of cement concretes with monomers or oligomers ensures the possibility of producing concrete polymers possessing high density and strength characteristics.

The above basic classification has been adopted in the international literature [46, 62]. Further, the following international abbreviations are commonly used at congresses, symposia, workshops, and in journal articles: PC—polymer concrete; PCC—polymer cement concrete; PSC—polymer sulphur concrete; and PIC—polymer impregnated concrete.

The workability of the polymer concrete, as in the case of cement concretes, is determined by the plasticity of the mix and is primarily dependent on the type and quantity of the synthetic resin used and also on the dispersability of the filler and the ratio between the fractions of the fillers and the aggregates. Polymer concrete mixes can be divided into four main groups based on their plasticity; their average compositions are given in Table 1.1. In this Table, the lower values of the quantity of resin refer to the heavy concretes and the higher values to the light concretes using porous aggregates.

The most important factor determining the properties of polymer concretes is the type of polymer binder and its compatibility and high adhesive bond

Table 1.1. Classification of polymer concrete mixes based on workability

| Material | Plasticity of polymer concrete mix depending on composition | | | |
	Plastic	Normal	Stiff	Dry
Aggregates	51–52	53–55	56–57	58–60
Sand	23–24	25–26	25–27	27–28
Mineral flour	10–12	10–11	8–9	6–7
Synthetic resin	12–15*	9–12	6–8.5	5–6

*The amount of hardener is selected depending on consumption of the resin.

with concretes based on inorganic binders. These are determined not only by the type of binder but also by the type of hardening system. For example, polymer concretes based on furan resins hardened by acidic catalysts should not be combined with cement concretes since, during hardening, the acid catalyst can damage the molecular structure of the cement concrete. For maximum possible elimination of such adverse phenomena, the classification of polymer concretes based on the types of polymer binder and the type of hardeners is given in Table 1.2.

Table 1.2. Classification of polymer concretes based on types of binders and hardeners

Sl. No.	Class of polymer concrete	Type of polymer concrete	Type of hardener	Synthetic resins	Hardeners
I	Furan	Furfural-acetone	Acidic	Furfural-acetone resins FA and FAM	Benzenesulphonic acid (BSA), sulphuric acid compound (SAC)
		Furan-phenol-formaldehyde	-do-	Furan-phenol-formaldehyde compound	-do-
		Furan-melamine	-do-	Furan-melamine compound	-do-
		Furyl	-do-	Furfuryl alcohol	BSA, ferric chloride
	Phenol	Phenol-formaldehyde	-do-	Phenol-formaldehyde resins SFZh-3032, SFZh-3016, etc.	BSA, water-soluble sulphoacids
	Carbamide	Carbamide	-do-	Urea-formaldehyde resin, KFZh, etc.	BSA, phosphoric acid, aniline hydrochloride

Sl. No.	Class of polymer concrete	Type of polymer concrete	Type of hardener	Synthetic resins	Hardeners
II	Acetone-formaldehyde	Acetone-formaldehyde	Alkaline	Acetone-formaldehyde resin ATsF-2 and ATsF-3	NaOH 25% and polyethylene polyamine (PEPA)
III	Polyester	Polyester maleinate	Peroxide type	Polyester resins PN-1, PN-3, PN-62, PNS-609, etc.	Initiators—peroxides and hydroperoxides; accelerators—cobalt naphthenate, etc.
		Polyester acrylate	-do-	Polyester resins MGF-9, TGM-3	-do-
	Vinyl	Vinyl	-do-	Monomer of methyl methacrylate MMA, etc.	-do-
IV	Epoxide	Epoxide ED	Amino compounds, etc.	Epoxide resins ED-16, ED-20, ED-22, etc.	Polyethylene polyamine (PEPA), UP-0633, etc.
		Epoxide EP	-do-	Epoxide-polyamide compound	
		Furan-epoxide	-do-	Furan-epoxide compound FAED	

An analysis of the most common aggressive media and degradation processes noticed in polymer concrete structures showed that all external influences can be divided into two main groups based on the type of action on the polymer material. The first type includes the action of water and some other liquid media causing a reversible adsorptive reduction of strength; the second type includes the effects causing irreversible degradation processes.

In turn, the effects causing irreversible degradation or deterioration processes in polymer material can be divided into three main subgroups. The degradation is caused by the dissolution of the polymer by different solvents; accelerated formation of free radicals and atoms and chemical action with the aggressive products associated with the deterioration of the polymer molecules.

The *first subgroup* represents the dissolution of the polymer binder by different solvents. For a binder based on thermosetting resins with a spatial cross-link, this type of action is manifest in limited ranges. There are highly active solvents for each type of binder.

For polymer concretes based on thermoplastic polymers, the intensity of the action of solvents can increase significantly; in such a case, the operational conditions and the types of permissible solvents acting on such polymer concretes should be more thoroughly studied.

The *second subgroup* combines the degradation processes associated with the action of heat and various types of radiation. It is known that sometimes the free radicals and atoms arising in the polymer under the action of heat, solar radiation, and radioactive radiations take active part in breaking down the polymers. Being reactive by nature, the free radicals and atoms break the polymer molecules and these fragments form new free radicals which also take part in the degradation process. As a result, the structure of the polymer, its chemical composition and molecular weight undergo change and, as a consequence, the physical and mechanical properties of the polymer also change.

Such degradation processes usually proceed most intensely in thermoplastic polymers with a linear structure. In this case, the breakdown of the polymers can be significantly inhibited by blocking the free radicals which arise initially as well as those formed as secondary products of the reactions. For these purposes, various stabilisers based on the derivatives of phenols, amines, sulphides, and organophosphorus compounds are added to the polymer at 0.1 to 3%.

The *third subgroup* represents the degradation of polymers associated with ionic and molecular processes which mostly occur under the influence of acids, oxidising agents and alkalis.

Thus, by classifying the diverse reactive media and the external processes according to their typical effects and generalising the results of laboratory investigations and the data of research on polymer concrete products and structures used in various industrial works and installations under the action of aggressive media, we can identify four main types of degradation of polymer concretes. Each type can be defined by the sum total of prominent features.

All reversible processes arising in polymer concretes under the action of water and other non-aggressive liquids can be placed in the *first type*. As a result of liquid diffusion deep into the material, a weight increase and adsorptive reduction of strength arise. On drying, i.e., evaporation of the liquid from the material, the initial weight and strength are practically restored.

Processes which cause the dissolution of the polymer and its washout from the polymer concrete structure can be placed in the *second type*. In this process, there is a weight reduction of the material and also its strength reduction.

Degradation phenomena of the polymer binder associated with the appearance of free radicals and atoms can be placed in the *third type*. This type of destruction causes brittleness of the material and strength reduction. In this process, there is no perceptible weight variation of the material.

The *fourth type* of degradation of the polymer binder is associated with the breakdown of polymer molecules under the action of acids, oxidising agents and alkalis. This type of deterioration causes an increase in weight, strength reduction, and spalling of the material (Table 1.3).

Table 1.3. Classification of the processes of degradation of polymer concretes

Nature of action	Type of degradation	Degradation processes	Evaluation of degradation process	Factors controlling degradation kinetics
Physical processes (saturation with liquid non-aggressive media)	I	Adsorption and reduction of surface energy of solid body	Weight increase of material and strength reduction	Diffusion rate and stressed state
Physical and chemical processes (action of solvents)	II	Dissolution of polymer binder	Weight reduction of material and strength reduction	Diffusion rate and stressed state
Physical processes (action of temperature, solar radiation and radioactive radiation)	III	Physical and chemical degradation	Increase of brittleness and strength reduction	Intensity of action
Chemical reaction of aggressive media with polymer binder	IV	Acidic, oxidative, and alkaline degradations	Increase of material weight, appearance of cracks, and strength reduction	Rate of diffusion, concentration of aggressive medium, stressed state, and temperature

In the case of reinforced polymer concretes falling under classes II and IV (according to the classification based on the type of binder and hardener (Table 1.2)), the steel reinforcement will find the most favourable conditions since the alkaline medium of the hardener promotes the passivation of steel while the high density of the polymer concrete protects it quite reliably from coming into contact with the aggressive medium. The conditions are least favourable for the reinforcement in polymer concretes hardened by various acids. The third group of polymer concretes occupies an intermediate position with respect to these characteristics (Table 1.4).

In practice, the degradation is not always of an entirely single type but, in most cases, the predominant action and the ancillary actions can be identified.

Table 1.4. Classification of polymer concretes according to type of hardener and stability of reinforcement

Hardener	Stability of reinforcement depending on type of fillers and aggregates			
	Acidic rocks (granites, etc.)	Neutral rocks (andesite etc.)	Porous	Carbon-graphite
Polymer concretes hardened by acids	Good	Good	Satisfactory	Unsatisfactory
Polymer concretes hardened by polyamines and alkalis	Good	Good	Good	Satisfactory
Polymer concretes hardened by peroxides	Good	Good	Good	Satisfactory

The reduction of intensity or the suppression of all types of degradation processes involves the correct selection of the polymer binder that is most stable under the action of actual aggressive media and achieving the maximum possible density of the polymer concretes so as to reduce the diffusive penetration of aggressive media deep into the material.

The classification of the types of polymer concrete degradation into the above four main types facilitates not only a more correct selection of the type of polymer concrete depending on the operational conditions and a more rational approach for formulating new polymer concrete compositions, but also helps to predict the behaviour of the reinforcement in a given situation.

2

Characteristic Features of the
Materials Used

2.1. Synthetic Resins, Monomers, Hardeners and Modifying Additives

Polymer concretes can be produced using practically any synthetic binder and also any filler or aggregate. However, considering the cost and availability of resources, requirements of density, strength, deformability, chemical resistance, and several other properties, only a comparatively small number (12 to 15 types) of polymer binders are used in the highly developed countries to produce polymer concretes. However, taking into consideration the various combinations between these monomers and oligomers and the possibilities of modifying the polymer binders of a given type with additives, the number of polymer binders used at present runs into several dozens.

Thus, in the USSR, for civil and industrial applications, furan, furan-epoxide, polyester, epoxide, phenol-formaldehyde, acetone-formaldehyde, carbamide, acrylic resins, monomers of the vinyl series, etc. are used as binders for various polymer concretes; polyester, epoxide, and phenol-formaldehyde binders based on acrylates and methacrylates are used in the Federal Republic of Germany; epoxides, polyesters, phenol-formaldehydes, vinyl esters, monomers of the vinyl series, etc. are used in the USA, England, France, Japan, and other countries.

Epoxide resins. Among the wide range of various types of epoxide resins, resins of type ED-16, ED-20, and ED-22 are mainly used in the production of polymer concretes. These resins are produced by the condensation of epichlorohydrin and diphenylol propane (bisphenol) in an alkaline medium. Generally, the resin is a transparent liquid of light yellow colour with a comparatively low viscosity. Polyethylene polyamine (PEPA), hexamethylene diamine, pyridine, UP-633, etc. are used as hardeners for these resins.

Epoxide resins possess a set of valuable properties: high mechanical and dielectric characteristics, high adhesion to most materials, good damping properties, low shrinkage, etc. However, the type of hardener and its quantum significantly influence their physical, mechanical, and chemical properties.

The possibility of modifying epoxide resins by various polymers makes for a wide variation of their properties; thus strength, thermal stability or elasticity can be improved as required. Thermosetting resins (phenol-formaldehyde and furan), thermoplastic materials (polyamides, polysulfides, etc.) are used most often with epoxide resins.

Of late, epoxy-urethane and epoxy-organosilicon compositions have come into use. Epoxy-urethane compositions contain epoxide groups and urethane fragments (polyisocyanate biureate and glycidol). Conventional hardeners, including aliphatic amines, can serve as hardening agents. These compositions possess high thermal resistance, improved dielectric properties and elasticity. The main properties of these compositions are:

Strength, MPa:
    compressive   130
    tensile   100
Relative ultimate elongation, %   15
Modulus of elasticity of elongation, MPa 2.9×10^3
Dielectric constant   3.8
Tangent of dielectric phase angle,
    tg δ, at frequency 10^6Hz 1.9×10^{-2}

The modification of epoxide oligomers by unsaturated organosilicon compounds and reactive organosilicon oligomers holds much promise [13]. On modification of ED-20 by unsaturated organosilicon compounds, which at the same time serve as diluents of epoxide resins, the strength characteristics and thermal resistance of the polymers increase considerably along with improved technological parameters (viscosity decreases and durability increases) (Table 2.1).

Table 2.1. Properties of modified polymers

Modifying agent	Strength, MPa			Relative ultimate elongation %
	Compressive	Bending (flexural)	Tensile	

CH$_3$ C$_2$H$_5$ CH$_3$
  │   │   │
HO–C–CH = CH–Si–CH = CH–C–OH
  │   •   │   │
CH$_3$ C$_2$H$_5$ CH$_3$
	220	193	101	3

CH$_3$ CH$_3$ CH$_3$
  │   │   │
HO–C–CH = CH–Si–CH = CH–C–OH
  │   │   │
CH$_3$ CH$_3$ CH$_3$
	260	245	120	3.3

Furan resins represent a group of compounds which contain the furan ring in their molecular structure. These are the products of condensation of furfural and furfuryl alcohol with phenols and ketones [1, 4]. Furan resins types FA and FAM are prominent among the large group of these compounds.

Properties of Furan Resins Types FA and FAM

	FA	FAM
Density, g/cm^3	1.14	1.137
Viscosity according to VZ–4, sec	25–40	24
Content, %, of		
water	0.6–0.8	0.4–0.5
furfural	1.1	1.2–1.3
dry residue	79–80	87–88
Acid number, mg KOH	8.07	17.7
pH of aqueous extract	4.2	4.65
Rate of gelatinisation at 120°C, sec	45–55	50–70

Several binders of the type of FAFF, possessing many favourable properties, resulted on combining furan products with phenol-formaldehyde, carbamide, organosilicon and other resins.

Polymer concretes based on furan resins possess high strength, wear and cold resistance, while their chemical resistance is generally excellent [29, 62, 75].

Furan-epoxide resins (FAED) represent compositions in which many characteristics surpass the properties of each of the constituent components individually. The physical and chemical properties and the dependence of the degree of polymerisation on the epoxide resin content and the hardening duration of the furan-epoxide compounds are shown in Tables 2.2 and 2.3 [29].

Table 2.2. Physical and chemical properties of furan-epoxide resins

Index	FAED-20	FAED-30	FAED-40	FAED-50	FAED-60	FAED-70	FAED-80
Content of epoxide resin ED-20, % by weight	20	30	40	50	60	70	80
Density, g/cm^3	1.141	1.143	1.146	1.15	1.152	1.156	1.16
Viscosity according to VZ–4, sec	28	44	114	127	183	246	305
Viscosity by Ostwald method, Pa × sec	97.9	223.8	351.2	561.4	1546.4	2500	–
Dry residue, % by weight	92.2	92.6	93.4	94.4	95.1	96.6	98.1
pH of aqueous extract	6.35	6.42	6.51	6.55	6.7	6.75	6.8
Rate of gelatinisation at 120°C with 20% PEPA, sec	127	123	120	94	70	66	43

Table 2.3. Hardening of Furan-Epoxide Resins

Type of composition	Degree of hardening (%) depending on duration of hardening (days) at room temperature					
	10	20	25	30	35	40
Epoxide resin ED-20	77.2	77.2	77.3	77.1	77.2	77.3
FAED-20	52.1	75.2	75.1	75.4	75	75.3
FAED-30	59.2	76.5	76.4	76.2	76.3	76.5
FAED-50	76	78.4	78	78.3	78.1	78.2
FAED-60	84	86.2	86.2	86.7	87.1	87.4

Note. After holding for one day at room temperature and heating for 24 hr at 100 to 110°C, the degree of polymerisation of FAED-10 and FAED-40 was 94 and 97% respectively.

The hardening of furan resins at normal temperature proceeds most completely by ion mechanism. Anhydrous aromatic sulphonic acids or sulphochlorides (toluenesulphonic acid, *p*-toluene sulphochloride, *p*-toluenesulphonic acid, benzenesulphonic acid, etc.), mineral acids such as sulphuric, phosphoric, hydrochloric, etc., and metallic chlorides such as ferric chloride, aluminum chloride, etc. can be used as hardeners for furan resins [62, 75].

Aromatic hardeners are usually crystalline materials and hence must be dissolved or fused before use. This poses certain technological difficulties. Further, sulphonic acids, being catalysts, do not react with the resins and being soluble in water, can be easily extracted, but this increases porosity and impairs physical and mechanical properties of the hardened polymer. In spite of these deficiencies, they are most extensively used for hardening furan resins.

Metallic chlorides are very expensive, in short supply and their reactivity is low.

Concentrated sulphuric acid is an extremely active hardener. The process of hardening of the compositions under the action of this acid proceeds extremely violently with large heat liberations, this being one of the main reasons for the impaired physical and mechanical properties of the hardened products.

The method of blocking the aromatic hardeners, especially benzenesulphonic acid, after hardening, by ion-exchange resins, is extremely interesting and holds promise. The addition of a comparatively small quantity of an ion-exchange resin of the amine type (2–4% of the weight of FA or FAM) reduces the amount of the benzenesulphonic acid extracted by roughly 25–30%.

Technical benzenesulphonic acid consists of monosulphonic acid of benzene 98.4–98.6%, free sulphuric acid H_2SO_4 1.2 to 1.4%, and benzene not more than 0.2%.

Studies carried out at the Reinforced Concrete Research Institute and other organizations in the USSR on the use of sulphuric acid diluted with water-soluble sulphoacids as hardener for furan resins showed that, at ratios of 1:1 to 1:2 of these components, the optimum amount of the hardener lay in the range 10 to

15%. In this case, the degree of polymerisation of the binder and the strength characteristics of polymer concretes based on FAM are not only not inferior to those of similar compositions hardened by benzenesulphonic acid, but are even superior in general. Sulphuric acid diluted with water-soluble sulphoacids is technologically easier to handle than benzenesulphonic acid; the heating operation becomes redundant while the possibility of crystallisation on cooling, which is characteristic of benzenesulphonic acid, is eliminated.

Acetone-formaldehyde resins (ATsF) represent products of polycondensation of acetone and formaldehyde at the molar ratio 1:2 or 1:3 in an alkaline medium. A 5% solution of caustic soda is used as a catalyst. It was established that these are low molecular compounds but complex in the composition of ketone alcohols [2].

Water-soluble ATsF resins should satisfy the following requirements:

	ATsF–2	ATsF–3
External appearance	Viscous homogeneous liquid, light yellow to yellow in colour	Viscous homogeneous liquid, colourless, or light yellow
Solubility in water	Unlimited	Unlimited
Content of dry residue (concentration), %	85–94	88–92
Content of hydroxyl groups, %	12–16	19–25
Density at 35°C, g/cm^3	1.25–1.3	1.22–1.26
Viscosity according to VZ–4, sec	60	60–80

Investigations showed that resin ATsF–2 is more suitable as a binder for polymer concretes [64]. The ATsF resins become infusible and insoluble on adding amines and caustic alkalis to the resin. A new, comparatively cheap binder of light colouration and hardened by alkaline products was thus produced. These properties helped produce polymer concretes with new, valuable properties in a wide colour range, possessing high stability to oils and other types of petroleum products, and solutions of salts and alkalis. Moreover, polymer concretes based on ATsF resins go well with the cement concretes due to the presence of alkaline hardeners in them.

On modifying the acetone-formaldehyde binder with phenolic alcohols, the physical and mechanical properties improve while acid resistance increases significantly.

Polyester resins. Polyester unsaturated resins are increasingly being used in civil and industrial applications; they can be divided, depending on type of compounds, into polyester maleinates and polyester acrylates.

(a) *Polyester maleinates* belong to the class of thermosetting polymers produced by the method of polycondensation. Unsaturated polyester maleinate resins are oligomers belonging to the class of heterochain polyesters, the complex ester group of which is invariably the structural element of the main polymer chain.

The resins of this series are the products of polycondensation of di- or polyfunctional acids and alcohols containing reactive double bonds between the carbon atoms [4, 14]. Their relative molecular weight does not exceed 1500 to 2000.

The ability of the polyester resins to harden at room temperature is explained by the presence of unsaturated bonds—the products of the first stage of polycondensation. The content of unsaturated groups in the polyester depends on the amount of maleic acid (or its anhydride).

Polyesters obtained by the reaction of maleic acid with glycol (polyethyleneglycol maleinate) are capable of polymerisation and co-polymerisation.

The hardening of unsaturated polyester resins proceeds as a result of co-polymerisation between an unsaturated polyester and a liquid monomer on heating or under the action of initiators and accelerators. Styrene is widely used as a monomer and methyl methacrylate to a lesser extent.

Co-polymerisation results in a compound of linear chainlets of the polyester with 'cross-linking bridges' formed by the monomer molecules. During this reaction, the polyester resin is hardened to a hard product with a spatial structure. The co-polymerisation reaction of the polyester with styrene under the action of initiating additives is accompanied by a significant exothermal effect.

The indigenous [USSR] industry produces over 15 types of polyester maleinate resins possessing diverse physical and mechanical properties. Of these, the resins listed in Table 2.4 are more frequently used in the production of polymer concretes.

Resin PN–3 is characterised by high thermal stability, PN–62 and PN–63 by low inflammability, PN–15 by high chemical stability and PNS–609–22 m by the absence of volatile solvents.

Polyester resins in most cases are hardened by using hardener-initiators: hydroperoxide of isopropyl benzene (hyperis) or methyl ethyl ketone and accelerators-activators, i.e., 10% of cobalt naphthenate solution in styrene or dimethyl aniline.

(b) *Polyester acrylates* are produced by simultaneous condensation of unsaturated dibasic acids with glycols, glycerine, or pentaerythrite in the presence of a monobasic unsaturated acid. Polyester acrylates TGM–3 and MGF–9 are more widely used in the indigenous industry.

Polyester acrylate TGM–3 is a product of the condensation of triethylene glycol and methacrylic acid. Polyester acrylate TGM–3 is produced in the form of a 96% solution in benzene with a density of 1.06–1.12 g/cm^3 and viscosity at 20°C 10–40×10^3 Pa-sec. Polyester acrylate is yellowish-brown in colour.

Polyester acrylate MGF–9 is produced by condensation of methacrylic acid, triethylene glycol and phthalic anhydride in an acidic medium followed by neutralisation of the product. It is sold in the form of a 96% solution in toluene of yellowish-brown colour with viscosity at 20°C 100–350×10^3 Pa-sec.

Table 2.4. Basic characteristics of some polyester maleinate resins

Index	PN-1	PN-3
Density at 20°C, g/cm³	1.12–1.15	1.12–1.15
Viscosity at 20°C:		
Pa-sec	—	—
according to VZ–1, sec	20–40	20–50
Durability at 20°C, min	60–120	60–180
Volume shrinkage, %	8.5–9	9–9.5
Ultimate bending strength, MPa	80–110	60–85
Specific impact strength, J/cm²	60–100	70–110
Modulus of transverse elasticity, MPa × 10³	2.2–2.8	2–2.5
Marten's yield temperature, °C	45–55	45–60

Index	PN-62	PNS-609-22 m
Density at 20°C, g/cm³	1.26–1.29	1.2–1.3
Viscosity at 20°C:		
Pa-sec	(1300–2000)10³	—
according to VZ–1, sec	—	150–200
Durability at 20°C, min	60–300	120–480
Volume shrinkage, %	9.5–10.5	8–9.5
Ultimate bending strength, MPa	50–70	50–60
Specific impact strength, J/cm²	30–50	30–70
Modulus of transverse elasticity, MPa × 10³	3–3.2	2.8–3.1
Marten's yield temperature, °C	55–60	50–70

The hardening of polyester acrylate resins is the result of co-polymerisation of linear polyesters with monomers together with homo-polymerisation of the resin constituents. The reaction proceeds by radical mechanism. The initiators are peroxides together with accelerators, the latter represented by cobalt and manganese naphthenates, tertiary amines, mercaptans, and other materials possessing reducing properties [14].

Benzoyl peroxide, cyclohexanone, methyl ethyl ketone, and hydroperoxide of isopropyl benzene are widely used for the hardening of polyester acrylates. The efficiency of these peroxides is best manifest when paired with some accelerators: benzoyl peroxide with dimethyl aniline, peroxide of cyclohexanone with cobalt naphthenate, hyperis with cobalt naphthenate, and so forth. It should be pointed out, however, that when using the above-listed systems for hardening polyester acrylates, the rate of the hardening reaction at normal temperature is extremely low and takes a few days, this being the main impediment to their practical application in polymer concretes. The addition of fillers and aggregates to the binder composition reduces even more the rate of polymerisation since, in this case, the heat of the polymerisation reaction is inadequate for auto-heating of the entire mass of the polymer concrete, i.e., there is no thermal initiation of the reaction.

A method was suggested [70] for hardening the unsaturated polyester acrylates at room temperatures in a comparatively short duration (2 to 3 hr) using the hardening system: hyperis-cobalt naphthenate, and hardening co-accelerator methyl vinyl Aerosil (silica powder).

Thiocol Hermetic U–30 m was used to enhance the elastic properties of the polyester acrylate binders.

The optimum composition of the binder based on polyester acrylate resin (% by weight) is as follows:

	Composition 1	Composition 2
Resin MGF–9	83	—
Resin TGM–3	—	83
Thiocol Hermetic U–30 m	10	10
Hyperis	2	2
Cobalt naphthenate	4	4
Methyl vinyl Aerosil	1	1

On introducing a third component, methyl vinyl Aerosil, which performs the role of a co-accelerator of the reaction, into the composition of the initiating system, the rate of polymerisation of polyester acrylate increases to such an extent that the polymer concretes based on them hardened at room temperature and gained up to 60% of maximum strength in a day. The co-accelerating role of methyl vinyl Aerosil is evidently explained by the presence of a large number of unsaturated bonds in its molecules; these bonds open up easily in the presence of peroxide compounds [70] and perform the role of primary active radicals intensifying the polymerisation process of unsaturated polyester acrylates.

The addition of methyl vinyl Aerosil to the binder containing Thiocol Hermetic U–30 m promotes better contact between the binder and the filler and the formation of a denser structure, resulting in high tensile and bending strengths.

Phenol-formaldehyde resins (FFS) represent products of polycondensation of phenol with formaldehyde in the presence of a catalyst.

The process of polycondensation of phenol with formaldehyde occurs as a result of the sum total of successive and parallel reactions of two types of compounds: polymerisation and polycondensation. Thermoplastic (novolacs) and thermosetting (resols) phenol-formaldehyde resins can thus be produced. The chain molecules of resonal resins consist of phenol nuclei joined with methylene groups ($-CH_2-$) or other bonds ($-CH_2-O-CH_2$). The molecular weight varies from 300 to 800.

Due to excess formaldehyde added to the reaction when producing resol resins, the molecules of these resins, unlike those of novolac resins, contain free methylene groups. The greater the content of methylene groups, the better the performance of the resin and its ability to further chemical transformations. Apart from methylene groups, the molecules of resonal resins contain free hydroxyl groups.

The presence of benzene nuclei in phenol-formaldehyde resins ensures their resistance to thermal degradation while the hydroxyl groups impart to them high adhesion to metals and non-metallic materials.

Investigations showed that polymer concretes possessing excellent physical and mechanical properties and high chemical resistance to several aggressive agents, can be obtained by using phenol-formaldehyde resins types SFZh–3032 and SFZh–40–KO.

Basic Properties of Phenol-formaldehyde Resins

	SFZh–3032	SFZh–40–KO
Density, g/cm^3	1.216	1.2
Viscosity according to VZ–4, sec	50	55
Content of dry residue, %, not less than	75	60
Content of free phenol, %, not more than	10	5
Content of free formaldehyde, %, not more than	5	4
Storage life, months, not less than	3	1.5
Cost, rubles/ton	440	330

The phenol-formaldehyde resins become infusible and insoluble on the addition of acidic catalysts to the resin. Various acids can be used as catalysts for phenol-formaldehyde resins. Good results were obtained in the case of polymer concretes based on phenol-formaldehyde resins types SFZh–3032 and SFZh–40–KO when using benzenesulphonic acid at 20% of the weight of the resin.

The hardening of phenol-formaldehyde resins by acid catalysts is not always favourable for the strength properties and the diffusion permeability of the hardened composition; a further danger is corrosion of the steel reinforcement used in the structure.

At present, compositions using acid-free hardeners, such as alkylamine, are known but their use requires the modification of phenol-formaldehyde resins by polyisocyanate [59] which greatly increases the cost of the compositions. When, however, sulphurous anhydride is used as an acid-free catalyst [60], the strength characteristics of many such compositions decrease markedly. Therefore, work on developing more effective acid-free hardeners is urgently called for.

Carbamide resins. Urea- and melamine-formaldehyde resins falling in the class of carbamide (amino-aldehyde) resins represent the products of simultaneous condensation of urea or melamine with formaldehyde in water or water-alcohol.

The carbamide resins become infusible and insoluble under the influence of acidic hardeners or heat together with hardening accelerators.

In terms of the volume of production, carbamide resins occupy one of the first places in many countries and, in terms of cost, urea-formaldehyde resins are the cheapest compared to the other types of synthetic resins. Apart from low cost, the other positive features of these resins are the low toxicity and more favourable working conditions when producing polymer concretes based on them. Such polymer concretes belong to the class of not readily inflammable materials.

The main drawback of phenol-formaldehyde and carbamide resins is the high content of water (30–40% of the weight of the mass). Therefore, the physical and mechanical characteristics of polymer concretes based on these resins depend largely on the effectiveness of binding the free water in the system. To realise strength and adequate stability of polymer concretes, a part of the water is chemically bound, thus bringing its content to a minimum. With this objective, phosphogypsum, gypsum, polyisocyanates, etc. are added to the composition of the polymer concrete.

Dozens of varieties of carbamide resins are produced in many countries (more than 15 types in the Soviet Union alone) but only two or three types are largely used for producing polymer concretes.

Main Properties of Carbamide Resins M19–62 and KF–Zh

	M19–62	KF-Zh
Density, g/cm^3	1.25–1.3	
Concentration (content of dry residue), %	65–70	65–70
Viscosity according to VZ–4, sec	40–90	40–80
Content of free formaldehyde, %	0.7–1.2	1–1.5
pH of the medium	7.2–8.5	7.5–9

The hardening catalysts for carbamide resins are organic (oxalic, citric, and acetic) and inorganic (sulphuric, hydrochloric, and phosphoric) acids and also some salts (ammonium and zinc chlorides). Experience has shown that aniline hydrochloride $C_6H_5NH_2HCl$, which dissolves well in water, and resin KF–Zh are the most suitable for building purposes. The catalyst is a greyish-green powder with a density of 1.222 g/cm^3, containing 98.9% aniline hydrochloride, 0.5–1.5% moisture and 0.1% insoluble matter. A drawback in using this catalyst is its comparatively high cost.

An extremely interesting new type of polymer concrete was produced by filling the carbamide resins with phosphogypsum and hydrolytic lignin: polymer binder—urea-formaldehyde resin, phosphogypsum, and natural polymer—hydrolytic lignin.

Phosphogypsum is produced as a by-product in the production of mineral fertilisers and contains about 2% phosphoric acid.

Hydrolytic lignin is a waste product of sulphuric acid treatment of wood and plant material for producing ethyl alcohol, furfural stock, edible and protein yeasts, and other products. Hydrolytic lignin contains 0.3–3% H_2SO_4 as also some complex organic acids. As a result of two simultaneously proceeding processes, i.e., hydration hardening of phosphogypsum and polycondensation of carbamide resin catalysed by lignin, a complex structure is formed in which the properties of the polymer binder, phosphogypsum and lignin mutually supplement and strengthen each other. Moreover, hydrolytic lignin is a dispersed-reinforcing filler whose addition greatly increases the strength of the mix. The limited amounts of organic and inorganic acids present in the phosphogypsum and lignin are adequate for hardening the system and no additional hardeners are needed.

Further, a significant part of the free water contained in the carbamide resin not only combines with the phosphogypsum, but is also absorbed by lignin, thus promoting the formation of a very dense structure and reducing shrinkage deformations.

Methyl methacrylate monomer MMA (methyl ester of methacrylic acid) is a transparent colourless liquid with a typical odour and very low viscosity. The content of the methyl ester of methacrylic acid in this product is not less than 99.7% and methacrylic acid not more than 0.2%. The presence of free polymer is not permitted.

Depending on the use of the polymer concrete, peroxides and hydroperoxides together with various amines can be used as hardeners for the methyl methacrylate system.

The main advantage of using these monomers as a binder is that low viscosity helps them hold a large amount of fillers and aggregates while remaining workable. The mix can be easily coloured as desired and can acquire a strength of over 30 MPa in 1.5–2 hr after production.

Investigations carried out at the Reinforced Concrete Research Institute in collaboration with the Institute of Physical Chemistry of the Academy of Sciences, USSR, and also by the State Construction Committee of the Turkmenian Soviet Socialist Republic on earthquake-proof constructions, helped in developing compositions of polymer slurries and polymer concretes called 'Elastocrils'. This term presently covers several compositions characterised by the system of hardening, types of fillers and aggregates, types of plasticisers and stabilisers and field of application.

For compositions of the type Elastocril-1, benzoyl peroxide (BP) with aromatic tertiary amine, i.e., dimethyl aniline (DMA), is used as a hardening agent. In this system, benzoyl peroxide plays the role of an initiator while dimethyl aniline functions as an accelerator for the dissociation of the initiator into free radicals. For Elastocril-2, the hardening system consists of hyperis, which plays the role of an initiator, and polyethylene polyamine, which acts as an accelerator.

For Elastocril-1, dispersed polymers type ABS (emulsion co-polymer of styrene with acrylonitrile and butadiene) or polystyrene is used as a stabiliser (thickener). For Elastocril-2, the stabilisers used are the low molecular butadiene-acrylonitrile latex type SKN-10-1A and others, as also ABS and polystyrene. To reduce the volatility of methyl methacrylate, a film-forming agent from the group of low molecular weight paraffins is dissolved in it in the course of production. Apart from the aforelisted components, plasticisers, dyes, and other constituents may figure in the composition of the binder.

Investigations showed that the binder based on monomer MMA hardens at room temperature by the mechanism of free-radical polymerisation with considerable exothermal heating of the mix.

It should be pointed out that compositions of the type Elastocril-1 harden well under water and can be applied on freshly laid cement concrete.

Average Compositions of Binder Based on MMA, % by Weight

	Elastocril-1	Elastocril-2
Methyl methacrylate MMA	100	100
Benzoyl peroxide (BP)	7	—
Hyperis	—	3
Dimethylaniline (DMA)	2	—
PEPA	—	3
Low molecular weight paraffin	0.5	0.5
ABS emulsion or polystyrene	10	—
Low molecular weight rubber	—	20–30
Dibutyl phthalate	3.5*	

*Dibutyl phthalate is used to produce a paste of benzoyl peroxide.

In the west (Federal Republic of Germany, USA, France, Italy, and others), unsaturated polyester resins, epoxide resins, methacrylate resins, and vinyl ester resins are presently preferred among the diverse resins used in the world for producing polymer concretes. This, however, does not take into consideration the production of moulding sands in the castings industry, which can also be regarded as a kind of polymer concrete.

Unsaturated polyester resins and their activators used in western countries. Unsaturated polyester resins (PN) represent solutions of unsaturated polyesters in reactive solvents. These are produced in the course of the polycondensation of unsaturated and saturated dicarboxylic (fatty) resins with glycols. By correctly selecting the basic constituents and using the required additives, a significant diversity of material properties is ensured with respect to their technological workability as well as the properties of the end products obtained from them.

The hardening of PN resins is effected by polymerisation, which is carried out by the addition of reagents (hardeners or, when required, catalysts). Further, the unsaturated groups of chain (linear) molecules of polyesters enter into the reaction with the unsaturated groups of solvents and combine with each other with the help of styrene bridge bonds as:

Liquid state

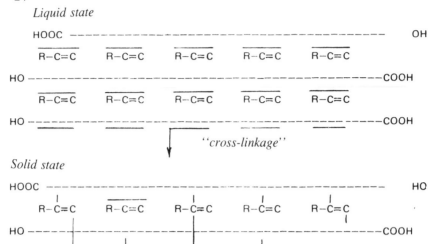

Solid state

From the initial liquid material, under conditions of heat liberation in the course of the reaction, a cross-linked 3-dimensional structure of infusible and insoluble plastic mass—thermosetting plastic—is formed. On transition from the liquid into the solid state, a volume shrinkage in the range of 5–8% occurs depending on the type of resin. (These parameters relate to the pure resin and not to the polymer concrete.)

The greater the double bonds (unsaturated groups) in the composition of the polyester molecules, the smaller the cells of the polymer grid cross-linked in the course of hardening and the greater the reactivity of the corresponding resin. In other words, a comparatively large amount of heat is liberated in the course of the reaction and the process of hardening itself proceeds quite rapidly. The finely cellular grid ensures an improved modulus of elasticity and hardness, thermal stability under load, and chemical resistance, but at the same time also causes some increased volume shrinkage. The polymers cross-linked into a grid of large cells have a low reactivity and a very low thermal stability under load (Martens or Vicks), but a very high elasticity and impact strength. In terms of reactivity, the resins are subdivided into rapidly hardening, highly reactive or medium reactive, and slow-hardening resins.

The unsaturated polyester resins can be roughly subdivided into the following categories according to their properties: *standard*—these correspond to the standard requirements with respect to technological workability in the hardened state and hence are used in many fields; *thermally resistant*—standard resins with improved thermal resistance and high chemical resistance; *not readily inflammable*—these possess in the hardened state (because of their chemical structure and also the presence of appropriate additives or supplements) a self-extinction property, i.e., burn only in the presence of an external source of fire; *resilient*—these have a low reactivity, possess good ultimate elongation and can be used independently as well as in admixture with other types of resins; these are very hard; *special types of resins*—these include, for example, air-drying resins and others; air-drying resins harden in the presence of atmospheric oxygen or on adding paraffin even at room temperature without tackiness; further, like chemically resistant resins, they possess high resistance to the action of water, acids and alkalis, but are of no practical importance in the production of polymer concretes.

When producing polymer concrete, the unsaturated polyester resins are usually hardened at room temperature by the addition of peroxides and accelerators. The addition of accelerators is absolutely essential to activate the peroxides at a very low temperature. The accelerator serves to replace the heat of thermal hardening, dissociates the peroxide in the course of the reduction reaction and hence the process is referred to as an oxidising-reducing system.

The following are the *systems hardening in the cold*: benzoyl peroxide with amine accelerator; peroxide of cyclohexane with cobalt accelerator; peroxide of methyl ethyl ketone with cobalt accelerator; peroxide of acetylacetone with cobalt accelerator; peroxide of ketone, perester, or hydroperoxide with vanadium accelerator.

The first four systems are used more often while the fifth is used only in special cases (system with vanadium accelerator) since its dosing should be particularly accurate and used in an extremely limited range.

Temperature is an important factor for hardening in the cold (cold hardening). Usually, cold hardening is carried out at temperatures up to 40°C, the optimum being the range from 15 to 30°C. At very low temperatures (around 0°C), the system benzoyl peroxide-amine works relatively well while systems with a cobalt accelerator provide a fairly slow, poor-quality hardening. This is due to the liberation of a very small amount of heat during polymerisation. Further, the components prepared using the system benzoyl peroxide-amine (which should conform to special requirements) should also be subjected to additional thermal treatment to ensure fairly good hardening. At temperatures above 30°C, however, additional acceleration resulting from the liberation of excessive heat reduces the material processing time, and hence the amount of the peroxide and the catalyst needs to be reduced. This, in turn, adversely influences the quality of hardening. In this situation, inhibitors can be used while the amounts of peroxide and catalyst remain unchanged (as at normal temperature); the addition of inhibitors is related to the ambient temperature and the required processing time with increase or decrease of the volume. The amount of the additive usually varies between 0.05 and 0.5%.

Epoxide systems. Epoxide resins are produced by converting the polynuclear phenols, mainly diphenylpropane (bisphenol A) and epichlorohydrin. The process of producing epoxide systems is represented as follows:

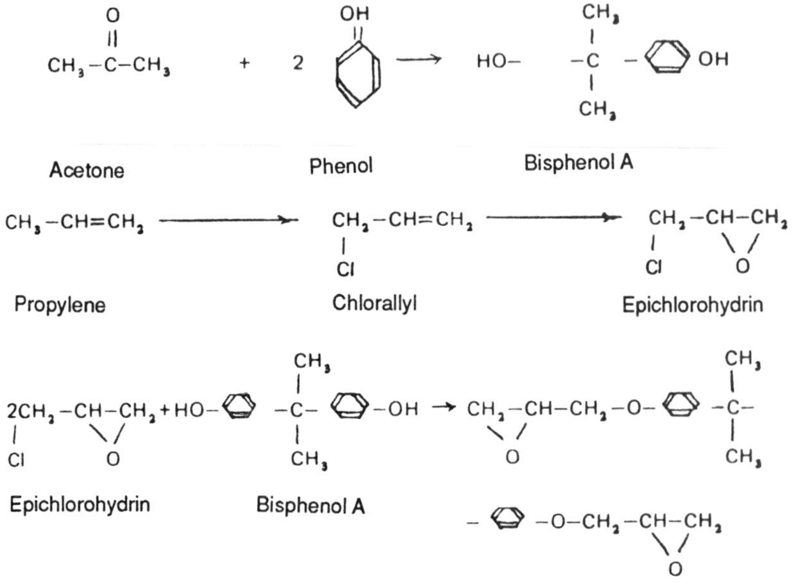

Acetone   Phenol   Bisphenol A

Propylene   Chlorallyl   Epichlorohydrin

Epichlorohydrin   Bisphenol A

Epoxide resin (liquid)

The epoxide resin is hardened in the course of producing the polymer concrete by cold hardening for which amine hardeners are ordinarily used. These hardeners enter into reaction even at temperatures of around 5°C. Hardening can be represented as a reaction of multistage polymerisation (polyadditives) without the reaction products dissociating during the formation of a 3-dimensional grid of polymer molecules.

Unlike unsaturated polyester resins to which the hardener is added in small amounts (1–4%), efforts are made to obtain stoichiometric amounts of resin and hardener when hardening the epoxide resin.

The required amount of the hardener may be calculated quite simply on the basis of equivalent weights[1]:

$$\left.\begin{array}{l}\text{amount of hardener}\\\text{per 100 parts resin}\end{array}\right\} = \frac{\text{equivalent weight of hydroamine}}{\text{equivalent weight of epoxide resin}} \times 100$$

It is desirable to select the optimum types of resin and hardener for making any binder material but, in the case of a system using the epoxide resins, the selection of the correct ratio of the resin and hardener is absolutely essential while keeping in view the qualitative as well as quantitative aspects. This is confirmed by the fact that the firms producing epoxide resins supply the resin and the hardener together while the suppliers of polyester or methacrylate resins do not produce hardeners.

----

[1] The indexes of the required equivalent weight are given in the technical documents of the raw material suppliers.

Methyl methacrylates and their activators. Acrylic or methacrylic acid represents the starting compound for monomers of methyl methacrylates. Only esters of these acids are used in polymer concretes. Further, their main component is methacryl methyl ester (Latin abbreviation MMA). MMA is produced on an industrial scale not by esterification of the acids, but as follows:

$$CH_3-C-OH_3$$
$$|$$
$$O$$

$$|\quad HCN$$

$$CH_3$$
$$|$$
$$CH_3-C-CH$$
$$|$$
$$OH$$

$$|\quad H_2SO_4$$

$$CH_3\quad \downarrow$$
$$CH_2-C-CO\,NH_2-H_2SO_4$$

$$H_2O\quad |\quad NH_3/H_2O \qquad\qquad CH_3OH/H_2O$$

$$CH_3 \qquad\qquad CH_3 \qquad\qquad\qquad CH_3$$
$$CH_2-C-COOH \quad CH_2-C-CO\,NH_2 \qquad CH_2-C-COOH$$

Methyl methacrylate resins are characterised by low viscosity which, for resins used in producing polymer concrete, is below 1000 Pa-sec, and possess fairly reliable technological workability even at relatively low atmospheric temperatures approaching the freezing point. They are inflammable at temperatures of about 10°C, which necessitates extreme care in storage and processing of these materials. Unlike in the production of other binder materials, for processing methacrylate resins, explosion-proof electrical equipment and machinery are usually used.

Methyl methacrylate resins are hardened by polymerisation effected as in the case of unsaturated polyester resins by the addition of accelerator (promoter) and catalyst. Benzoyl peroxide of 50% concentration is exclusively used as a catalyst (when producing polymer concretes). The catalyst is supplied as a powder, liquid, or paste. The amount of catalyst used is 0.5 to 5% by weight and the optimum amount is determined relative to the ambient temperature. The amount of hardener has practically no influence on the quality of the polymer concrete.

An amine is used as a promoter for activating the hardening process. In many cases, the amine is supplied by the manufacturers in admixture with the resin and not as a separate constituent. In this case, an adverse factor is the impossibility of establishing the required amount of the accelerator by the factor' processing the resin. Thus, the time of hardening depends wholly and completely on the external temperature.

Two methods are essentially distinguished in the processing of methyl methacrylate resins: liquid-powder system, also called monomer-polymer system (or abbreviated as 'mo-po system'), and cast resins.

When adopting the method using the liquid-powder system, the liquid consists of a monomer mixture and accelerator. The powder component of the system contains, apart from inorganic fillers, polymer powder and powder catalyst. When mixing the constituents, the polymer powder initially swells under the action of the solvent and is later dissolved. Thus, the polymer concrete composition contains a highly viscous monomer-polymer solution that hardens later during polymerisation.

When using the system of cast resins, the polymerisation product dissolves immediately in the monomer mixture, thereby eliminating the addition of polymers to the powder constituent. The monomer-polymer liquid (containing the promoter) is mixed only with inorganic fillers and hardens as a result of adding the catalyst.

When processing the resins manually, it is preferable to use a liquid-powder system (keeping in view its technological workability) even though the expense of procuring the required raw material is undoubtedly high. During mechanical processing, the method of cast resins is favourable not only with respect to price, but also because of another advantage: as a result of the presence of dissolved polymers in the system, the shrinkage of the finished structural component caused by the varying densities of the polymer and monomer system is significantly reduced.

Vinyl ester resins and their activators. Vinyl ester resins are produced by the chemical reaction of an unsaturated organic acid (for example, methacrylic acid) with epoxide resins. The unsaturated polymer chains are later 'cross-linked' by radical (radical-chain) polymerisation with styrene as the monomer, as in the case of producing standard polyester resins.

Compared with other types of resins, vinyl ester resins are characterised by unusual stability throughout the pH range. They can be used, for example, even in the presence of 37% hydrochloric acid or 50% caustic soda solution. Among all other resins used for producing polymer concrete or plastics reinforced with fibreglass, vinyl ester resins show excellent resistance to bleaching media and can be used at places where they come into contact with chlorine, chlorine dioxide, hydrocarbons, etc.

Vinyl ester resins are stored and processed in the same manner as described above in the case of polyester resins. The activators used are also the same as in the case of polyester resins.

It is known that compatibility of the molecular properties of the surfaces of the filler and synthetic resin is a necessary condition for intensifying the action of the highly dispersed fillers added to the polymer. At the same time, most of

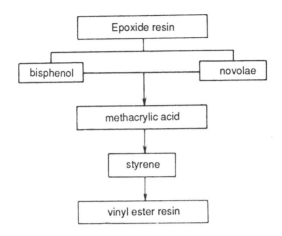

the mineral fillers are usually hydrophilic and cannot fully exhibit their activity, as reflected in improved physical and mechanical properties.

In turn, the cohesive strength of the polymer and the adhesive strength of the system 'polymer-filler' depend on the nature of molecular interaction. Further, ionic and covalent interactions lead to the formation of chemical bonds (with energy 250–330 kJ/mole) while hydroxyl, carboxyl and similar other groups form the so-called hydrogen bonds (with energy up to 33–42 kJ/mole). It is natural that the strength of filler polymer compositions is maximised by the formation of chemical bonds between the molecules of the polymer as also at the boundary of separation of the polymer-filler phases. Consequently, for hydrophilic fillers, processes of adsorption modification are of particular importance, i.e., the modification of the molecular properties of the particle surfaces of such fillers by controlled adsorption of diphilic long-chained surface-active agents (surfactants).[1]

The specific properties of surface-active agents arise from the typical arrangement of their molecules. The molecules of most surfactants consist of two main portions, i.e., hydrophilic polar group and hydrophobic non-polar radical. On preferential adsorption on the surface of phase separation, such molecules modify the surface activity of the mineral filler, as a result of which the adhesion of the polymer to the filler increases and the plasticity of the resin increases, leading to some reduction in binder consumption while maintaining or even bettering the physical and mechanical properties of polymer concretes.

---

[1] The innumerable works of P.A. Rebinder and his school have demonstrated that the surfactants represent a powerful and effective means of modifying the surface activity of the fillers.

For mineral fillers with an acidic property (andesite, quartz flour, etc.), cation-active surfactants are more effective while anion-active surfactants are suitable for basic fillers (dolomite, diabase, etc.). Further, the effect is significant in a comparatively narrow range of concentrations. Excess of the surfactant sharply impairs all the properties of the composition.

Investigations carried out at the Institute of Physical Chemistry, Academy of Sciences, USSR, Reinforced Concrete Research Institute and other organisations showed that for modifying the surface of silica fillers in compositions based on polyester resins, good results were obtained by using cation-active surfactants of the type octadecyl amine (ODA) and Alkamone OS-2 (ALK). Octadecyl amine and Alkamone were less effective when used in polymer compositions based on furan resins FAM.

A study of the effect of various types of surfactants in polymer compositions based on furan resins showed that cation-active substances of the type catapin are more active.

Thus, octadecyl amine or Alkamone (OS-2) can be recommended for compositions based on polyester resins and catapin for compositions based on furan resins.

2.2. Mineral Fillers and Aggregates

It is known that mineral fillers and aggregates exert considerable influence on the properties of filler polymers and especially on the properties of polymer concretes since they serve as inert additives reducing the cost of the products, restrict the temperature and shrinkage deformations, and control the density, strength, hardness, and other physical and mechanical properties of the products.

For polymer concretes used for technical purposes, the types of mineral fillers and aggregates, their chemical and mineralogical properties, and also size composition have been adequately described [29, 70, 75, 88]. However, it is desirable to dwell briefly on the possibility of using industrial wastes in polymer concrete compositions. Thus, instances are known of using crushed wastes of polystyrene, polyvinyl chloride, polypropylene, and other polymer articles as aggregates in polymer concretes. About 10% of domestic wastes, i.e., used glass, could be recirculated. This, however, is generally expensive as it involves grading the glass according to colour and separating the stones and foreign impurities. Ground glass provides a fine-grained inert aggregate that could be used effectively in producing polymer concrete. In fact, many countries are already utilising this waste product.

The same could be said of vulcanised rubber wastes (old tyres) which also contain impurities. Ground rubber wastes could be mixed with synthetic resins and subjected to further processing. In actuality, however, such material is not always fit for use in polymer concrete even though it, too, is based on a synthetic binder.

2.3. Reinforcing Materials

Given their high strength characteristics (high bending and tensile strength) polymer concretes can be used with no reinforcement whatsoever. Nevertheless, some areas of application of polymer concrete do require its reinforcement to enhance the strength characteristics or to ensure high reliability (safety factor).

In principle, polymer concretes based on polyesters, epoxides, vinyl esters, MMA, and FAED can be reinforced in the same manner as concrete using a cement binder. Any problem that might arise would pertain only to the shrinkage of the polymer concrete. Steel and fibreglass are ordinarily used for reinforcement. Other fibres, such as carbon filament, are used only in special cases, which fall outside the scope of this book.

When using steel reinforcement, particular attention should be paid to the fact that the reinforcement *per se* cannot prevent shrinkage of the material, resulting in the deformation of the finished building structures. It should also be borne in mind that polymer concrete does not chemically combine with steel. The 'bond' between steel and polymer concrete is an adhesive one and hence for the reinforcement of polymer concrete, clean rust-free steel subjected to additional milling should be used. Steel fibre would ensure additional reliability.

Table 2.5. Comparison of the properties of high-strength combination profiles with different materials

Index	Steel wire	Aluminium wire	Polyamide fibre	Combination profiles (68% fibre + glass)
Tensile strength, MPa	1800	360–460	500	1500
Yield point in elongation, MPa	1600	280–390	200	—
Relative ultimate elongation, %	5	4.5–7	—	3
Modulus E	21×10^4	7×10^4	3×10^4	5.5×10^4
Length of fracture, km	23	12–17	50	75
Density, g/cm^3	7.85	2.7	1.14	2

Fibreglass is used in different forms: mats, cut fibres, or rods. The use of mats or cut fibres reflects the current level of engineering development. However, mention should be made of rods, designated as high-quality combination profiles, recently developed by a West German firm. This new combination material represents a semi-finished product using synthetic materials as hardeners and fibres as reinforcement in a controlled orientation (in space). As a result of the high fibre content (about 85% by weight), the optimum orientation of the fibre and their air-free impregnation, the material possesses high strength and rigidity, improved static and dynamic properties, excellent electrical insulation properties, and corrosion resistance at low density.

Using a new method of coiling, combination profiles possessing several additional advantages are being produced: smooth surface, geometrically true section, improved transverse strength, low sensitivity to stress concentrations, as also (because of the resin-covered surface) improved technological workability and atmospheric stability (Table 2.5).

3

Theory of Structure Formation using Polymer Concretes and Physical and Chemical Principles in Formulating their Compositions

3.1. Material Models of Polymer Concretes

Many authors hold that the material model of polymer concretes should be defined in terms of an overall successive dependence of 'constituents—structure—properties—application'. In most cases, however, only mathematical models of 'constituents—properties' have been developed. Nevertheless, such models should be regarded as an important step from science to materials technology or a bridge between theory and practice.

The Institute of Reinforced Concrete Research, Prague Institute of Engineering [8], Warsaw Technical University [25, 26, 42], Nihon University, Japan [56], and individual scientists are presently working on the development of more complete material models of polymer concretes. Thus, L. Czarnecki [25, 26] has emphasised the need for developing material models as a prerequisite for rational formulation of polymer concrete mixes. Manson [48] has said they are necessary for a better understanding of the principles of formulating polymer concrete mixes; according to the Institute of Reinforced Concrete Research, such models are essential for further advancement in the field of polymer concretes.

Several reports presented at the IV International Congress on Polymers in Concrete (Dramstadt, Federal Republic of Germany) by K. Konrad, V.V. Paturoev, Zh. Slivinskii, Kh. Shori, and others pointed out that the material model continues to remain one of the basic scientific problems.

On the whole, the properties of polymer concretes should be regarded as derived from adhesion and energy reactions between their constituents [42, 62]. Then, the mechanical properties, especially the modulus of elasticity E, can be calculated with the help of an adhesion mechanism (bulk model) by using various modifications and rules of mixing. Y. Ohama predicted in this manner the compressive strength of polyester polymer concrete based on the application of the rule of mixtures and an appropriate empirical linear equation [56]. R. Bares [8] made a significant advance by introducing into the model, apart from other refinements, an 'internal surface' without which the structural model cannot be regarded as adequately representative. Further, it is necessary to remember that pores as a special type of filler represent an important structural factor whether they have been incorporated in the structure or not.

The major drawback of the bulk model is its limited applicability only at the macrostructure level. This model does not take into consideration the effect of energy reactions between the constituents of polymer concrete.

Commencing from the end of the 1960s and in the early 1970s, several scientists recognised the significant influence and importance of the morphology of supramolecular structures and the orientation mechanism on the surface of phase separation in the system 'resin–filler'.

3.2. Morphology of Supramolecular Structures

Aspects associated with the effect of fillers on the rate of build-up and re-laxation of shrinkage stresses, strength and deformation characteristics of highly filled polymer compositions have been discussed in great detail [62] and the formation mechanism of supramolecular structures which determines the above properties has been demonstrated. However, for a better understanding of the material presented below, the basic assumptions of the principles identified are discussed taking into consideration the latest experimental data.

The investigations carried out in collaboration with the Institute of Physical Chemistry, Academy of Sciences, USSR [45, 78, 88, 89] have established that the rate of polymerisation and relaxation processes in polymer compositions based on oligomers is determined by the effect of fillers and the distribution of active groups in the polymer. The nature of supramolecular structures arising during polymerisation is determined by the distribution of active groups in the system.

The effect of fillers as nuclei of structure formation on the morphology of supramolecular structures and on the physical and mechanical properties was first determined for polyester resins type PN-1. Quartz flour with a specific surface of 0.8 m^2/g (as determined by low-temperature adsorption of nitrogen) was used as a nucleus-forming agent. A polarisation microscope MP-7 was used to study the structure.

It was shown [78, 88] that, at a low concentration of the filler (10%), an oriented and highly stressed structure is detected around the filler particles. The size of the polymer shell surrounding the filler greatly exceeds that of the filler grains and goes up to 50–80 μm. Such structures are transparent in polarised light and represent centres of concentration of internal stresses. On high magnification ($\times$ 1800), filler grains are seen inside the polymer shell. The colour of the polymer structure around the filler grains is uneven and changes along the section of the filler surface. It was established that this colour change is due to the uneven distribution of internal stresses. Further, maximum stresses were detected in the layers adjoining the filler surface.

With the growing number of structure-forming centres, the size of the oriented polymer structures decreases and approaches that of the filler particles.

The structural morphology identified suggests that the reactive functional groups of the secondary supramolecular structures and not of the individual macromolecules react with the filler surface.

When studying the supramolecular structures of polyester compositions in an electron microscope, titanium dioxide (rutile form) with a specific surface of 10 m^2/g [62, 87] was used as filler. For the study, carbon-platinum replicas were prepared by the oxygen etching method.

The largest supramolecular structures of circular form with a diameter of some microns were detected when the polyester resin contained 10% rutile filler (Fig. 3.1, left). A densely packed globular structure found at the centre of such formations is transformed in the subsequent layers into an anisodiametric structure oriented perpendicular to the filler surface. The anisodiametric structure probably arises as a result of the opening up of the globules.

As the proportion of the filler increases up to 50% (Fig. 3.1, right), the oriented anisodiametric structure predominates. This increases adhesion, improves the rigidity, and, as a consequence, increases shrinkage stresses.

Fig. 3.1. convincingly demonstrates the nature of variations of supramolecular formations in relation to the degree of mineral fillers present in the polyester resins. The size and extent of orderliness of the secondary supramolecular structures arising around the filler particles differ greatly from the molecular structures of unfilled polymers or in a polymer body at considerable distance away from the filler surface.

The variation patterns detected in the morphology of supramolecular structures were found applicable in the case of furan resins also.

More detailed investigations on several oligomer systems of polyester, furan, and other resins helped to identify the main variation patterns of physical and mechanical properties and the supramolecular structure of materials based on these resins in the presence of fillers. It was established that the fillers most frequently used in polymer compositions (quartz flour, andesite, marshalite, Aerosil, and many others) are active in relation to these resins and are capable of entering into specific reactions with them to form covalent or hydrogen bonds and also supramolecular structures that are more stable than in the polymer body. As a result, such filler systems are invariably damaged in the polymer body while the polymer shell around the filler particles is preserved [62, 87, 88].

The variation patterns of properties are common to oligomer polymer systems in the presence of active fillers and depend significantly on the concentration of the filler.

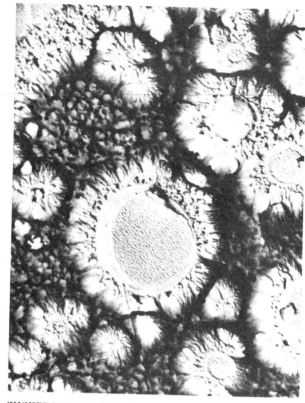

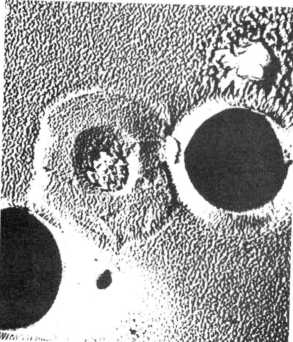

Fig. 3.1. Variation in the supramolecular structures of polyester compositions in relation to the rutile content (× 1500); left 10% and right 50% filler.

Further, it was found that globular structures predominate and, depending on the sizes and degree of orderliness, they exert a vital influence on the physical and mechanical properties of polymer compositions.

Studies on supramolecular structures of compositions based on furan resins confirmed the earlier results that the polymerisation and polycondensation reactions proceed through the stage of formation of the globular structure. This type of structure formation is preserved on adding various proportions of active fillers: only the size of the structural elements varies sharply. For example, a cross-linked structure of the largest structural elements arises on adding 10% andesite filler. Subsequent increase of the filler to 50 and 100% leads to a reduction of the structural elements. At 50% of andesite, the size of the supramolecular formations approaches the original particle size of the unfilled resin while, at 100% filler content, the size of these particles becomes smaller than in the original resin.

The reduction of the size of supramolecular structures is accompanied by an increase of their rigidity and, as a consequence, by an increase of internal shrinkage stresses and the rate of their build-up.

With an increase of moulding temperature to 80°C, the globular formations in the system have a much smaller diameter than in the case of hardening at 20°C. The size reduction of the structural elements in the course of hardening at 80°C is again accompanied by an increase in rigidity and shrinkage stresses as compared to systems hardened at 20°C.

A different type of structure formation is noticed in samples of FAM compositions filled with graphite which, as explained below, greatly reduces the internal shrinkage stresses when the content in the system is 100% or more.

The hardening of furan resin FAM in the presence of graphite with a distinct lamellar form is accompanied by the orientation of large structural elements along the tiny filler particles and the formation of cross-linked structure whose density increases as the amount of the filler increases. The lamellar structure of the graphite particles gives rise to the formation of a fibrillar structure close to its surface. This structure is oriented along the surface of the filler particles and leads to parallel disposition of fibrils relative to each other and in the polymer layers away from the filler surface. Further, it should be pointed out that the sizes of the supramolecular structures (length of fibrils) are 10 to 20 times more than in the compositions produced by using active fillers. Unlike the active fillers, graphite particles do not form centres of formation of secondary supramolecular structures and promote only a parallel disposition of the structural elements of the polymer relative to each other.

The adhesive bonds of furan resins with the graphite surface, which are much less compared to the active fillers and the fibrillar and hence the more elastic supramolecular structure, make for comparatively low shrinkage stresses in such compositions.

Based on the above premises, a conclusion of extremely practical importance can be drawn. On adding certain quantities of graphite flour to polymer concrete compositions containing active fillers, internal stresses can be significantly reduced.

The results of experimental tests showed that the addition of 5% graphite flour to polymer concretes containing quartz or andesite flour as active fillers reduces internal stresses by 25–30% while preserving the original strength characteristics.

Investigations demonstrated that a fibrillar structure can be produced even in the case of polymer compositions containing active fillers by treating the filler surface with surface active agents or by directly adding such agents to the polymer binder. At optimal concentration of the surface-active agent, the globular structure noticed before is transformed into a more orderly fibrillar structure with distinct anisodiametric dimensions. Such a structure leads to a significant reduction of internal stresses in the system while maintaining high adhesion and strength of the filled composition.

Thus, filled compositions of oligomer systems can be strengthened by forming in them a homogeneous orderly supramolecular structure which ensures partial relaxation of internal stresses to the initial level even during polymerisation. The investigations carried out convincingly demonstrate that the physical, mechanical, and chemical properties of polymer compositions can be modified over a wide range by controlled modification of the morphology of supramolecular structures and the degree of completeness of the spatial cross-linking of the polymer binder.

These and many other researches carried out at the Institute of Physical Chemistry, Academy of Sciences, USSR (P.I. Zubov, L.A. Sukhareva, and N.I. Morozova) established the hitherto unknown phenomenon of thixotropic reduction of internal stresses in filled polymer systems which was registered as a discovery (Certificate of Discovery No. 190). The scientific importance of this discovery lies in that it provides an essentially new approach to understanding the mechanism of generation and development of internal stresses and suggest practical methods of reducing them. It also refutes the generally accepted concept that intensifying the reaction between the structural elements of the polymer system should lead only to a build-up of internal stresses. The practical importance of this discovery lies in that it helps reduce objectively the internal stresses when formulating new polymer materials, including polymer coatings, adhesives, and other filled systems.

3.3. Theory of Structure Formation in Polymer Concretes

As pointed out already, the basic properties of polymer concretes are determined not only by the type of synthetic binder but also by the type of fillers and aggregates, their size composition, and no less importantly, by the correct selection of the proportions of various fractions.

As a result of the integrated fundamental investigations carried out at the Reinforced Concrete Research Institute in collaboration with the Institute of Physical Chemistry, Academy of Sciences, USSR, a general theory of structure formation in polymer concretes was developed based on various polymer binders [43, 44, 49, 89].

The developed theory is essentially based on commonly accepted principles and posits that the basic physical and chemical reactions of the polymer binder occur at the boundary with the surface of the finely dispersed filler while the nature of this reaction obeys the rule of extreme values, i.e., the extreme values of the resultant properties correspond to the optimum polymer content of the system (Fig. 3.2).

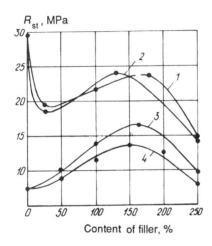

Fig. 3.2. Dependence of ultimate strength (R_{st}) of polymer compositions on filler content:

1—compositions based on quartz flour; 2—polyester compositions based on andesite; 3—FA compositions based on andesite; 4—FA compositions based on graphite.

The principles established enable depiction of the structural model of polymer concrete which should include the microstructure of the binder composition, mesostructure of the polymer solution, and the macrostructure of the system as a whole.

It has been experimentally established that the optimum properties of the binder compositions (polymer and the finely dispersed filler) can be realised at filler dispersions in the range 3000–5000 cm^2/g. The optimum polymer capacity is developed practically when the packing of the fillers is most dense. Further, at the boundary of phase separation, orderly oriented structures of the polymer are formed. These structures consist of supramolecular formations of the globular or fibrillar type which determine the basic properties of the microstructure of the binder composition depending on the nature of the polymer—filler bonds and supramolecular morphology.

The dependence of the strength of the binder on the dispersion of the filler, $R_b = f(S)$, in the practical range of dispersions, is also characterised by the presence of extremal values. As the dispersion increases to 0.3–0.4 m²/g, strength increases sharply reaching the maximum. Later, as a result of the self-adhesion of the particles and the impossibility of uniform mixing, a reduction in the strength of the binder is noticed.

A fairly complete idea of the kinetics of structure formation and the charac-teristics of the ultimate properties of the compositions studied, can be had from a simultaneous and comprehensive study of the morphology of supramolecu-lar transformations, processes of shrinkage, thermal expansion as a result of exothermal polymerisation and the corresponding internal stresses.

Studies on the morphology of supramolecular formations in the microstruc-ture of polymer concretes based on various types of polymer binders and on internal stresses, revealed a common pattern of the variation of supramolecu-lar formations depending on the polymer—filler ratio, chemical composition of the monomer or oligomer, and the nature of the filler. Simultaneously, methods were developed for controlled modification of the morphology of supramolecu-lar structures and the corresponding shrinkage by modifying the polymer binder or filler by surface active agents. It should further be pointed out that most surface-active agents are not only modifiers controlling the nature of forma-tion of supramolecular structures and reducing the shrinkage stresses, but are concomitantly good plasticisers.

Based on the above considerations, a scientific approach to formulating polymer concrete compositions can be represented in the form of a series of successive stages (Fig. 3.3).

The results are given below of studies on the basic patterns of structure formation which are responsible for the most important properties of polymer concretes and also on the development of an experimental—theoretical method for selecting the optimum compositions of polymer concretes.

The theoretical principles of selecting the polymer concrete compositions are based on the conditions of achieving maximum density and minimum con-sumption of the synthetic binder while maintaining high values of strength and other physical and mechanical properties of the material. The following assump-tions have been made. Since the specific surface of the aggregates in relation to the specific surface of the finely dispersed fraction of fillers in most cases is 1 to 2%, it may be assumed that the initial formation of the polymer concrete structure proceeds in two stages. Initially, the finely dispersed fraction of the filler mixes with the synthetic resin and forms an adhesive composition. It fills uniformly the intergranular space, envelops the much larger grains of the aggre-

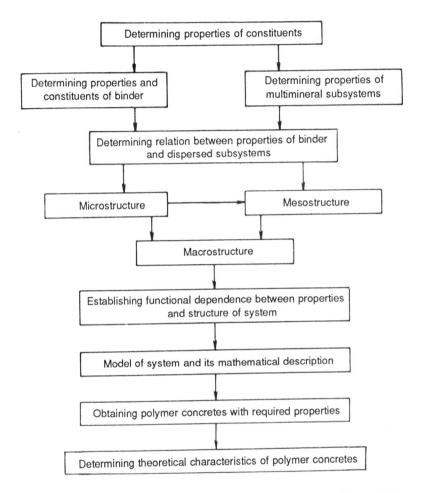

Fig. 3.3. Successive stages in formulating optimum multicomponent systems for a particular type of polymer concrete.

gate, and binds them into a strong impermeable structure of polymer slurry or polymer concrete.

If the strength of the aggregate grains exceeds that of the adhesive composition, the short-term and long-term strengths, deformability, and other physical and mechanical properties of the polymer concrete are determined by adhesive and cohesive bonds of this composition. Otherwise, the strength of the polymer concrete is evaluated from the strength of the weakest grains of the aggregate.

Thus, three structures should be distinguished in polymer concretes: microstructure of the adhesive composition, mesostructure of the polymer solution and the macrostructure of the system as a whole. Depending on the morphology of the supramolecular formations, the microstructure determines the rheological characteristics of polymer concretes. The properties of meso- and macrostructures are determined by the packing density of the aggregates and the presence of defects.

The proposed structural model of polymer concrete has been well acknowledged by specialists as an aid in studying the basic physical and mechanical properties and their variation patterns in very simple systems, i.e., compositions; later, having determined the optimum parameters and characteristics, these can be improved in more complex systems, i.e., polymer slurry and polymer concrete. Moreover, such an approach helps in a very deep study of the essentially complex physical and chemical reactions occurring during structure formation.

Studies on adhesive compositions are also important when they are used as anticorrosive coatings or synthetic adhesives.

The structure and properties of filled polymers (inclusive of adhesive compositions) as heterogeneous multicomponent systems generally depend on two factors [49]. The first is the very principle of producing filled materials by adding fillers to monomers or oligomers. These fillers differ in physical and chemical structure, size and form of the particles and their content in the system. The second is the result of changes in the physical properties and structures of the polymer matrix caused by the reaction at the boundary of separation of the polymer—solid body. The overall change in the properties of a filled system, compared to the original polymer, occurs as a result of the simultaneous action of both these factors and is therefore not additive. However, in all cases, the most important condition intensifying the action of fillers in filled systems is the adhesion of polymers to the filler surface and hence the nature of the bonds at the boundary of separation of the polymer—solid body. The chemical or physical reaction of the polymer with the filler surface determines the deformability of the filled polymer, the nature of stress concentrations on the filler particle surfaces, and the disintegration conditions of the filler.

The structure and properties also depend on the interphase reactions. These depend on the phase and physical state of the polymer, flexibility of its bonds, and the density of cross-links, and hence will differ for linear, amorphous, crystalline, and cross-linked polymers. At the same time, the primary factor that influences the main properties is the adsorption reaction of the polymer with the surface of the solid body which determines the adhesion, the number of possible conformations of bonds in the boundary layer and the limitations imposed by the geometry of the surface on the conformational set of polymer bonds [10].

As a result of the adsorption reaction, a boundary layer with altered properties is formed close to the filler particles. The thickness of such a layer varies widely depending on the nature of the polymer and the surface of the filler. With the increasing distance away from the surface, the properties of the boundary layer change steadily and no distinct boundary can be seen between the surface layer and the polymer in the body. The scouring of this boundary is seen in the absence of a sharp change in the physical properties on gradual transition from one zone to another.

It is well known that the properties of polymers are determined by the types of structural units which take part in a given physical process. For example, the effects associated with the mobility of segments are seen in the boundary layer to a larger extent than the effects caused by the reaction of supramolecular structures and their elements with the surface. Thus, the effect of the surface can be short- or long-range.

In general, it can be said that the long-range effect of the surface is the result of its penetration through macromolecules fully or partly bound to the surface by adsorption to the more distant layers by intermolecular reaction forces. Thus, this effect is a function of two factors: surface energy of the filler and the cohesive energy of the polymer. Apart from the adsorption reaction with the surface, the effects associated with a change in the conformation of polymer bonds, i.e., their flexibility, also exert a marked influence on the properties of the boundary layers.

The investigations carried out at the Institute of Physical Chemistry, USSR, in collaboration with the Reinforced Concrete Research Institute showed that, for linear and cross-linked amorphous polymers, the main result of the adsorption reaction and the conformational limitations imposed by the surface is the change in the molecular mobility of segments of bonds and side groups. Moreover, the distribution of polymer density close to the filler particles and at different distances from them will also vary.

The properties of the filled polymer also depend significantly on the technological conditions of production including the hardening temperature. Restraining the molecular mobility inhibits the course of relaxation processes during the formation of filled compositions and promotes the formation of a structure with less equilibrium. Some changes in the properties of the boundary layers can be explained by this uneven variation. However, there is adequate justification to assume that many properties of filled polymers are in equilibrium to the extent that they can be regarded as representing a state of equilibrium in amorphous polymers. This is particularly true of the microstructure of polymer concrete binder in which the extent of filling corresponds to the optimum polymer content.

The effect of fillers on the morphology of molecular formations in amorphous-linear or cross-linked polymers is far more difficult to investigate as the types of supramolecular structures formed are much less diverse compared to the filled crystalline polymers. Investigations of the sizes of globules and the nature of their concentration at different distances from the surface have shown that the size distribution of globules and their concentration are significantly modified in relation to most of the factors discussed above. Further, the effect of the surface on supramolecular formations extends much greater distances from the surface than were assumed before; in some cases, these distances can go up to 150–200 μm.

The surface properties of the layers which play an important role in the physical and mechanical properties of filled polymer compositions can be evaluated by using modern physical methods of research. An experimental study of molecular as well as supramolecular structures is possible at present.

According to the data of the Institute of Physical Chemistry, Academy of Sciences, USSR, more complete information on the effect of the filler on the structure of polymer compositions can be obtained by studying dielectric relaxation, which helps in determining the molecular mobility of different sections of bonds at different temperatures; applying the nuclear magnetic resonance method with respect to the size and temperature dependence of the second moment or the method of spin echo with respect to the temperature dependence of the duration of spin—lattice relaxation; employing the molecular probe method which helps evaluate the change and distribution of polymer density in the filled system and some other methods.

The first two methods provide information on the mobility variation of macromolecules under the influence of the filler surface and help in calculating the corresponding activation energies.

The molecular probe method is based on determining the shift ΔV according to the frequency of peaking of the luminescence spectrum under the influence of density variation ΔP of the polymer medium. The dependence $\Delta V = f(\Delta P)$ is first experimentally determined for a given luminophore before it is introduced into the polymer. By varying the thickness of the polymer layer between the filler particles, data can be obtained on the polymer density in the surface layers of different thicknesses.

When studying the swelling of filled specimens in a solvent in which the polymer swells only to a limited extent, it is possible to judge whether the density of linear polymers increases during their reaction with the filler consequent on the formation of additional units of physical networks or the effect of loosening up of the packing predominates.

For filled cross-linked polymers, the change in the effective density of the polymer network can be judged from the data of swelling.

Thus, for amorphous polymers, apart from the limited direct structural methods, there are practically no special methods for studying the structure and properties directly of the surface layers at the boundary of phase separation. Therefore, the development of more reliable and standard methods of studying the boundary layers is an urgent task. With the growing perfection of research methods, the data on the origin of additional levels of heterogeneity in the filled systems and on the morphology of supramolecular formations in the boundary layers, all of which represent the consequence of interphase reactions, will naturally be rendered more accurate and re-examined. However, independent of the methods adopted, the most important task for further research is the development of a method for quantitative evaluation of the heterogeneity of filled compositions, methods for modifying it objectively, and the development of a common theory of structure formation in such systems.

It should further be pointed out that the filler surface can catalyse or inhibit the hardening process of the binder and emerge as a chemical or selective sorbent and restrict the mobility of the polymer molecules or exhibit other properties. All these lead to a change in the hardening rate of the oligomer in the interphase layer, wettability of the filler by the binder, strength of the adsorption layer, and the formation of a structure that differs from that of the polymer present outside the zone of influence of the filler.

Studies on FAM compositions based on furan resins showed that on adding various fillers, the density of the binder decreased in the course of the first 15–45 min compared to the original value. Later, a definite equilibrium was established depending on the surface energy of the filler and the nature of the adsoprtion bonds. The density of the binder in the equilibrium adsorption layer was more than the value in the body. The loosening up of the oligomer phase in the initial period of contact with the filler is associated in all probability with the disintegration and rearrangement of the polymer binder structure under the influence of energy reactions of macromolecules with the filler surface [73].

With the increasing volume of the filler, the role of surface phenomena at the boundary of phase separation becomes increasingly more prominent since an increasingly greater amount of the polymer transits into the interphase surface layer.

Naturally, the reinforcing effect has a limit corresponding to the maximum polymer capacity when the entire binder or most of it enters the structured state.

In the zone of supercritical values of the filler, i.e., beyond the maximum polymer capacity, the binder is not adequate for completely wetting the surface of the filler particles and the strength of the composition falls again.

It has been shown [78] that the hardening processes of furan resins proceed through the stage of formation of supramolecular structures of the globular type with diameters ranging from 20–30 nm with the liberation of some amount of water which promotes the formation of a large number of defects in the polymer body. Therefore, the hardened resin is characterised by high brittleness and low tensile strength, thus pointing to a weak intermolecular reaction between the large globular elements of such a structure.

Active fillers promote the formation of an orderly supramolecular structure from anisodiametric structural elements in the course of hardening. Such a structure is characterised by very high strength—more than 2 or 3 times the tensile strength of the unfilled resins (Table 3.1).

In laboratory investigations, the nature of adhesion bonds in many cases is judged from the change in the strength indices of the material in relation to the thickness variation of the polymer film between the filler particles. This evaluation is not quite correct but does help in evaluating the basic principles of

Table 3.1. Dependence of adhesion (A) and tensile strength (R_t) of FAM composition on filler concentration

Volume concentration of the filler, %	Andesite		Graphite	
	Adhesion (A), MPa	Tensile strength, (R_t), MPa	Adhesion (A), MPa	Tensile strength (R_t), MPa
0	2.8	5.3	2.8	5.3
20	3.2	9.1	—	—
30	4.5	15.7	2.5	6.8
50	11.4	11.2	2.1	11.4
70	—	—	1.8	9

the influence of fillers on the physical and mechanical properties of the polymer compositions.

3.4. General Principles of Build-up of Shrinkage Stresses in Polymer Concretes

Internal shrinkage stresses arising during the formation of polymer concretes and their build-up during subsequent use represent one of the most important criteria for determining the long-term strength of these materials.[1]

As an example, Fig. 3.4 shows a specimen produced from unfilled furfural acetone resin FA which disintegrated spontaneously under the influence of shrinkage stresses and the action of water over a comparatively brief duration. The nature of the disintegration (regular spherical hemispheres) is interesting. It convincingly shows that shrinkage stresses attain maximum values and cause spontaneous disintegration of the material on slight weakening of the strength on the specimen surface due to the plasticising action of water. The spherical form of disintegration strikingly points to the nature of distribution of shrinkage stresses along the section.

The internal stresses in polymer compositions are associated with the phase transformation of the composition from a liquid to a solid in the course of hardening and incomplete relaxation processes. These are caused by several factors, such as shrinkage phenomena as a result of the proximation of oligomer molecules in the course of polymerisation and transformation from liquid into a solid state; formation of rigid supramolecular structures of the polymer and their high adhesive bonding with the filler particles and shrinkage due to the loss of volatile constituents.

---

[1] Investigations on polymer concretes, of theoretical and practical importance, were carried out for the first time by the author in collaboration with the Institute of Physical Chemistry, Academy of Sciences, USSR.

Fig. 3.4. Spontaneous disintegration of specimens produced from furfural acetone resin FAM under the action of shrinkage stresses following brief exposure to the action of water.

The progress of internal shrinkage stresses is usually expressed in the form of typical curves (Fig. 3.5) in which two sections can be distinguished. The first corresponds to the advance of internal shrinkage stresses under the influence of the above factors while the second reflects the nature of reduction of internal shrinkage stresses associated with the course of relaxation processes under conditions of stable external influences.

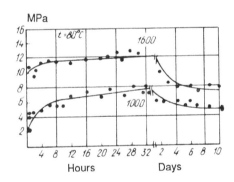

Fig. 3.5. Kinetics of the development of shrinkage stresses in polyester binder. Numbers on curves show the thickness of the polymer film, μm.

Thus, internal shrinkage stresses can be tentatively classified as temporary stresses (first section) whose action is manifest from a few hours to several days and residual, long-acting stresses (second section). Temporary shrinkage stresses are quite considerable and exceed in some cases the strength of the synthetic binder. These stresses are extremely dangerous as they can lead to micro- and macrofissures, i.e., disintegration of the impermeability of an article or structure. The residual stresses are usually considerably less than the temporary stresses. The danger from the former stresses lies in their prolonged action.

Investigations [44, 78] showed that the hardening of the polymer compositions in most cases proceeds roughly as follows. A small number of cross-links is formed during the early polymerisation process when the composition is still sufficiently elastic and relaxation processes proceed easily in it while internal stresses are practically absent. Depending on further cross-links, the number of cross-bonds and the rigidity of the composition increase and finally a moment sets in when a product with a highly dense spatial structure is formed. By this moment, shrinkage deformations and temporary internal stresses reach maximum values.

At the same time, it should be pointed out that the absolute value of shrinkage does not represent a criterion of internal stresses. At high shrinkage and low modulus of elasticity, internal stresses will be insignificant. A low shrinkage in materials with a high modulus of elasticity gives rise to significant internal stresses.

This picture is even more complex in the case of filled polymer compositions. An increase in the extent of filling of the system with quartz flour, andesite, marshalite and many other mineral fillers leads to a significant reduction of the shrinkage of the polymer composition. If it is assumed that shrinkage stresses depend exclusively on shrinkage deformations, the addition of mineral fillers should lead to a sharp reduction of shrinkage stresses. However, such an assumption does not take into consideration the corresponding increase of the modulus of elasticity and the more rigid adhesion bonds because of the formation of orderly supramolecular structures.

Investigations have shown that, as the extent of filling is increased, the modulus of elasticity increases far more rapidly than shrinkage decreases (Table 3.2). For example, for filled systems containing 300% by weight of the filler, shrinkage decreases roughly by 2 times while the modulus of elasticity increases by 4 to 5 times compared to the unfilled systems.

The maximum critical values of internal shrinkage stresses relative to the type and quantity of mineral fillers are shown in Table 3.3 which reveals that the addition of mineral fillers, irrespective of their type, to thermosetting resins significantly reduces the shrinkage of the polymer composition. At the same time, fillers such as quartz and andesite flour are highly active and capable of specific reaction with the binder by accelerating polymerisation and cause a significant build-up of shrinkage stresses. Graphite flour, an inert filler, inhibits polymerisation and reduces shrinkage stresses in the system while simultaneously reducing adhesion.

Dependence of the Adhesion of Furan Coatings on
Graphite Content in the System

Graphite content, %	0	50	100
Adhesion, MPa	14–16	10.5–12	8–9

Table 3.2. Effect of extent of filling with quartz flour on shrinkage, modulus of elasticity and internal stresses in polymer compositions

Type of binder	Quartz flour filler, %	Compressive strength, MPa	Volume shrinkage, %	Shrinkage reduction, %	Modulus of elasticity, MPa $\times 10^4$	Increase in modulus of elasticity, %	Internal stresses, MPa
PN–1	—	125	9	100	2.4	100	—
	50	116	7.75	116	2.3+	137.5	—
	100	137	7	128.5	5.3	221	—
	200	138.5	5.3	170	8.1	337.5	—
	300	134	4.5	200	10.5	437.5	—
	400	131	4	222	11.5	480	—
	600	76	—	—	7.3	304	—
FAM	—	142	7	100	3.1	100	3
	50	146	6	116	4.5	145	3.8
	100	160	5.4	127	7.1	230	5.5
	200	148	4.1	175	10.5	340	10.6
	300	132	3.6	194	13.7	442	15.2
	400	115	3.2	219	16.7	539	–
	600	—	—	—	9.5	306.4	–

+ *Sic*; probably should read 3.3—Technical Editor.

Table 3.3. Effect of various fillers on maximum critical internal shrinkage stresses of compositions, MPa

Filler	Content of filler, %			
	0	50	100	200
	FAM resin			
Quartz flour	2.8–3	3.8	5.5	10.6
Andesite flour	2.8–3	3.2–3.5	4.5	11–11.5
Graphite	2.8–3	2.5–2.8	2.1–2.5	—
	PN-1 resin			
Quartz flour	2.5–2.7	3.5	4.8	9
Andesite flour	2.5–2.7	3.2	4.5	9.6
Electrode graphite (flour)	2.5–2.7	2.1	1	—

Thus, one of the criteria for producing high-strength, reliable and durable polymer compositions is to find possible methods for reducing temporary and residual shrinkage stresses.

Innumerable investigations have shown that an effective and comparatively simple method of reducing the internal shrinkage stresses is the addition of surface-active agents to the polymer composition. An optimum amount of the surface-active agent not only reduces the internal shrinkage stresses signifi-

50

cantly, but also increases strength (Fig. 3.6). The addition of comparatively small amounts of cation-active surface-active agents to polymer compositions reduces the temporary internal stresses by 4 or 5 times and increases tensile strength by 30 to 60% (Table 3.4). The most active surface-active agent for the compositions under study is Katapin B-300 while Katapin BPV is less active. Alkamon OS-2 occupies an intermediate position. Since the composition of polymer concrete is highly complex, its compressive strength on adding the optimum amount of B-300 increases less significantly by 25 to 30%.

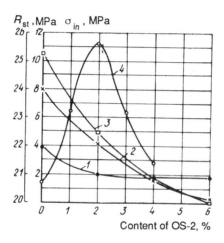

Fig. 3.6. Dependence of internal stresses (σ_{in}) and tensile strength (R_{st}) of polyester compositions on Alkamon OS-2 content:

1, 2, 3—internal stresses at 25, 60, and 100 parts by weight of quartz flour respectively; 4—tensile strength at 50 parts by weight of quartz flour.

It should be pointed out that studies on the effect of surface-active agents on internal shrinkage stresses did not pay adequate attention to these products from the viewpoint of their effect on the plasticising properties of the polymer mixture during its preparation. Recent studies showed that many surface-active agents act as good plasticisers for polymer concrete mixtures too.

Experimental studies of polymer concretes helped to establish the general principles of the development of overall shrinkage stresses during hardening, to determine their values in relation to the compositions of polymer concretes, and to make recommendations for reducing shrinkage stresses while maintaining or even improving their physical and mechanical properties. These studies resulted in a general theoretical equation for determining the shrinkage stresses

Table 3.4. Effect of type and quantity of cation-active surface-active agent on tensile strength of polymer compositions

Surface-active agent	Amount of surface-active agent, %	Ultimate tensile strength, MPa, of composition based on	
		FAM	PN-1
Katapin B-300	—	11.8	23
	0.25	14.3	27
	0.5	17	32
	1	18.3	31
	1.5	—	—
Katapin BPV	—	10.4	22.9
	0.25	11.2	—
	0.5	11.3	28.3
	1	13.8	26.5
	1.5	7.4	21.5
Alkamon OS-2	—	10	—
	0.5	13.2	—
	0.75	15.2	—
	1	11.2	—
	1.5	8	—

in reinforced polymer concretes:

$$\sigma_s = \frac{\epsilon_s(1 - e^{-\beta\tau})E_{p.c}\psi_1}{(1 - \mu)(1 + m)10^2}, \qquad \ldots(3.1)$$

where ϵ_s is the shrinkage of polymer concretes in time τ, %; β a coefficient of proportionality, l/h ($\beta = 0.055$ for polymer concretes based on polyester and furan resins); τ the duration of hardening, h; $E_{p.c}$ the modulus of elasticity of polymer concrete in time τ; ψ_1 the coefficient of relaxation of shrinkage stresses, $\psi_1 = 0.4$ to 0.6; and μ Poisson's ratio, $\mu = 0.24$ to 0.32;

$$m = E_{p.c}S_{p.c}/E_{st}S_{st} = (E_{p.c}/E_{st})(1/\nu)10^2, \qquad \ldots(3.2)$$

where $S_{p.c}$ and S_{st} are the areas of cross-section of the polymer concrete and the reinforcement, cm^2; E_{st} the modulus of elasticity of steel, MPa; and ν the coefficient of reinforcement of the structure, %.

For polymer concretes based on polyester resins, the maximum shrinkage is 3 to 4 mm/m and for polymer concretes based on FAM, 1 to 2 mm/m. In this case, for articles and structures in which the degree of hardening is closer to the maximum possible value, the following equations can be used to determine the shrinkage stresses with an accuracy adequate for practical purposes:

$$\sigma_s = \frac{\epsilon_{s-max}E_{p.c.}\psi_1}{(1 - \mu)(1 + m)10^2}. \qquad \ldots(3.3)$$

3.5. Formulation of Polymer Concrete Mixes

Cylindrical, defect-free specimens and standard panels were cast from polymer mixes and their tensile strength determined. While maintaining a given level of the binder, the amount of the filler possessing an identical specific surface was altered. Tests showed that initially, with an increasing amount of filler, the tensile strength rises gradually to maximum but falls subsequently. The optimum thickness of the binder film around the filler grains corresponds to the maximum strength of the mix (see Fig. 3.2).

It was also established that the overall dependence of the strength of the polymer mix on the filler dispersiveness is characterised by its maximum value ranging from 2500 to 3000 cm^2/g.

The values shown in Fig. 3.2 reveal that with the increasing amount of filler to a certain level, the average density, porosity, and compressive and bending strengths vary between extreme values although the experimental values of these characteristics are distributed in a fairly narrow range.

Thus, a definite structure and optimum physical and mechanical properties correspond to the optimum content of the filler in the system, which is called the *rule of experimental values*. It follows from this rule that, by determining two or three characteristics of the material of a correctly selected mix, it can confidently be assumed that the remaining characteristics will also have optimum values.

The researches carried out and an analysis of the results reported by other authors helped to draw very important conclusions: each type of thermosetting synthetic resin and mineral filler is characterised by an exact level of the filler, which ensures maximum strength of the composition; the reduction or increase of the filler leads to a significant reduction of the strength of the system; and the optimum amount of the filler is determined not only by the nature of the synthetic resin and filler, but also by the dispersiveness of the latter.

By knowing the pattern of strength variation relative to the degree of the filler, the viscosity of the synthetic resin and the specific surface of the filler, a mathematical dependence can be derived between these values and a theoretical equation arrived at for determining the maximum required amount of the binder for formulating a given composition.

Based on the experimental data, the film thickness can be calculated using the following equation:

$$\delta = V_b \eta_t / S_f m_f, \qquad \qquad \dots (3.4)$$

where δ is the thickness of the binder film, cm; V_b the volume of the binder, cm^3; S_f the specific surface area of the filler, cm^2/g; m_f the weight of the filler, g; η_t the ratio of the actual viscosity of the resin to the viscosity of the same resin taken as standard (20 sec by the VZ–4 method).

After transformations, the equation assumes a form that is more convenient for calculation:

$$\delta = m_b \eta_t / S_{sp} m_f \rho_b, \qquad \qquad \dots (3.5)$$

where m_b is the weight of the binder, g; and ρ_b the density of the binder, g/cm^3.

Calculations show that for optimum compositions of polyester, furan and other compositions, the effective film thickness of the binder around each grain of the filler is 1.5–2 μm while the total film thickness between the grains in an impermeable composition varies in the range 3–4 μm.

The consumption of the binder in the composition is then calculated using the following equation:

$$G_{comp} = (S_f m_f \rho_b \delta \eta_t) 10^{-3}, \qquad \ldots (3.6)$$

where G_{comp} is the optimum composition of the binder in the composition, kg; S_f the specific surface area of the filler (mineral flour), cm^2/kg; m_f the weight of the filler, kg; ρ_b the density of the binder, kg/dm^3; and δ the thickness of the binder film, $\delta \approx 0.00015$ cm.

Calculations of the amount of binder using equation (3.6) show that for oligomers, the ratio binder: filler = 1:1.5 to 1:2. Compositions containing such amounts of the filler possess maximum strength and rigidity.

For known size composition of dry aggregate mixtures, the optimum amount of the binder for polymer concrete is determined using the following equation:

$$G_{p.c} = [K(S_1 m_1 + S_2 m_2 + S_3 m_3 + S_n m_n) \rho_b \delta \eta_t] 10^{-3}, \qquad \ldots (3.7)$$

where $G_{p.c}$ is the optimum amount of the binder for polymer concrete of the adopted composition, kg: S_1, S_2, and S_3 the specific surface areas of aggregates of different fractions (coarse rubble, small-sized rubble, and sand), cm^2/kg; m_1, m_2, and m_3 the weights of the aggregates of different fractions, kg; and K a coefficient allowing for the increased amount of the binder required for separating the aggregate grains by the composition, $K = 1.05$.

Thus, by knowing the specific surface of the fillers and aggregates, the amount of the synthetic binder for polymer concretes of optimum composition is determined using equation (3.7). This equation, however, cannot establish in each given case the grain size of the aggregate and the optimum ratio between the individual fractions.

There are essentially two different methods for selecting the size composition of dense mixtures of multicomponent systems: by intermittent and continuous granulometry. For selecting the compositions of cement concretes, the first method is not widely used because of some complexity in determining the discontinuity of the fractions and the additional expenses involved in sieving the rubble and sand. Judging from the results of investigations of several authors, there is no advantage of using mixes with intermittent granulometry to justify these additional expenses.

Yet, since even an insignificant reduction in the consumption of polymer binder which can yield superdense mixes of agglomerates results in a significant reduction in the cost of the polymer concretes, it is economically advantageous to formulate compositions using intermittent granulometry.

Investigations have shown that when producing mixes using intermittent granulometry, the main difficulty of sieving the finely dispersed fractions and determining their optimum quantity in the mixture can be eliminated. This is because the difference in the sizes of the subsequent and preceding grains does not run beyond the size limits of the standard set of sieves while the effect of compaction, compared with the effect of compacting unsieved compositions, is so insignificant that there is no point in such a size separation. Hence a method of selecting dense mixes of polymer concretes using 'semi-intermittent granulometry' was adopted by us, which provides for size separation of stone aggregates alone and selection of the corresponding module of sand size. Finely ground additives have continuous granulometry, i.e., they are used without sieving. This method was used to select dense mixes of polymer concretes with synthetic resin consumption from 7.5 to 8.5% (165–185 kg/m^3) by weight of fillers and aggregates (Tables 3.5* and 3.6) while the earlier consumption of the binder, before using the method, ranged from 220 to 240 and in some cases even up to 320 kg/m^3.

Based on the investigations carried out, an experimental-theoretical method of selecting optimum mixes of polymer concretes was worked out, which incorporates the following features. The polymer concrete mixes are planned in three stages: the optimum composition of the binder is determined experimentally at first and the size of the rubble for the concrete mix with semi-intermittent granulometry and the amount of the fractions and their ratio are then calculated theoretically. After this, the aggregate composition is rendered more accurate in the device for selecting the dry mix.

3.6. Structural Strength of Polymer Concretes

The adopted model of polymer concretes, representing a complex composition consisting of micro-, meso-, and macrostructures, is the first attempt at developing the general principles and theoretical equations of the structural strength of such multicomponent systems. Further, the binder (polymer binder) represents the microstructure of the polymer concrete composition and is the basic and most important constituent of such a system.

Studies on physical and chemical processes of structure formation in direct experiments showed that the optimum microstructure of the binder depends on several properties of the monomers or oligomers used and primarily on their viscosity, adhesive capacity, and adhesion with mineral fillers, and also on the dispersiveness of the fillers, their form and content in the system.

Based on these premises, the structural strength of the polymer binder depends, in general, directly on the following factors:

$$R_b = R(G_v, A, S, P, \eta), \qquad \qquad \ldots (3.8)$$

---

* Omitted in Russian original or Tables misnumbered —Translator.

Table 3.6. Compositions of FAM polymer concrete selected by the method of semi-intermittent granulometry of Aggregates, %

Constituent	Theoretical		Rendered accurate in device for selecting dry mix	
	No. 1	No. 2	No. 3	No. 4
aggregate fraction, %:				
30	52	52	52	52
20	15.75	18	10	11
6	2.75	3.75	7	9
Sand, size 0.5 to				
2 mm	10	11.25	9.5	12
Andesite flour	11	—	12	—
Ground graphite	—	5.5	—	6
Furfural acetone				
resin FAM	7	7.5	7.5	8.5
Benzenesulphonic acid	2	2	2	2

Table 3.7. Results of compression tests of polymer concretes

No. of composition as in Table 3.5	Consumption of synthetic resin, %	No. of samples	Mean arith-metical strength, MPa	Mean square deviation	Variation coefficient	Degree of uniformity, K_u
1	7	15	86.4	+ 9.6	1.3	0.625
2	7.5	15	88.6	+ 9.83	1.25	0.61
3	7.5	15	82.3	+ 10.4	1.45	0.565
4	8.5	15	83.8	+ 9.7	1.32	0.604

where G_v is the ratio of the polymer to the filler; A the adhesion of the polymer to the filler; S the dispersiveness of the filler; P the porosity of the binder composition; and η the viscosity of the initial monomer or oligomer.

Modern mathematics helps to express with a high degree of accuracy the effective vital dependence of structural strength on polymer binder using theoretical equations. However, the use of such equations, even with computers, is practically impossible because of the many difficulties associated with the need to obtain all the fairly reliable characteristics and the corresponding transitional coefficients.

At the same time, from the developed theory of structure formation of polymer compositions and the proposed method of selecting optimum compositions, the principle of selecting the microstructure of polymer binder is based on the experimental determination of the optimum ratio of the actual constituents, i.e., a definite monomer or oligomer and the corresponding filler. In such an approach, it is possible to obtain the maximum possible strength R_b for a given system.

Having obtained the actual structural strength of the polymer binder and having collected sufficiently large and statistically processed data on the strength characteristics of polymer concretes, it is comparatively easy to derive a mathematical expression suitable for calculating the structural strength of the polymer concrete, $R_{p.c}$:

$$R_{p.c} = (R_b + K_1 R_{ag}) - (K_2 \sigma_{p.c} + R_p), \qquad \ldots (3.9)$$

where R_b is the strength of the binder; R_{ag} the strength of the aggregate; $\sigma_{p.c}$ the shrinkage stresses in the polymer concrete; R_p the loss of strength depending on the porosity; and K_1 and K_2 coefficients.

In equation (3.9), the strength of the aggregate R_{ag}, binder R_b, and internal shrinkage stresses $\sigma_{p.c}$ are determined experimentally while the loss of strength depending on the porosity of the polymer concrete R_p and coefficients K_1 and K_2 are determined by statistical processing of a large number of test results on different types of polymer concretes.

The practical application of equation (3.9) for calculating the structural strength of polymer concretes showed comparatively good agreement between the theoretical and experimental results.

Determination of the optimum polymer concrete mixes based on the principles of physical and chemical reactions between the constituents and the densest packing of fillers and aggregates helped to produce the most economic dense compositions with the minimum possible consumption of the synthetic binder yet possessing high chemical resistance. It was concomitantly noted that, depending on the purpose and the operational conditions, the physical and mechanical properties, requirements of the polymer concretes vary widely. The above method does not permit calculating the compositions of polymer concretes with pre-fixed strength and other characteristics however. Since polymer concretes represent multicomponent systems and all the constituents, to some extent or the other, mutually influence their ultimate strength, the determination and forecast of the latter by the usual methods is extremely complicated. For such systems, the change of strength in relation to composition should be regarded as an interrelated multifactorial process.

When studying and selecting polymer concrete mixes, mathematical methods of planning the experiments were therefore used and appropriate programmes developed for computers. The use of such methods provides more complete and reliable information while greatly reducing the experimental work [83]. Considering that the polymer concrete mixes should be selected using the optimum composition of the binder, the percentage contents of the resin and filler were adopted as constant.

The results of calculations using mathematical models on computers helped to obtain polymer concrete compositions using various binders for different operational conditions. The compositions of polymer concretes possessing maximum strength obtained using the mathematical methods of planning the experiments agree practically with the compositions obtained based on the theory of dense packing of fillers and aggregates and the minimum permissible amount of binder. The nomograms designed for equivalent yield help to predict the strength of different polymer concrete mixes and to formulate compositions with the required strength.

4

Physical and Mechanical Properties of Polymer Concretes

As already pointed out, depending on the type of synthetic binder, fillers and aggregates used, and the methods of production, polymer concretes may possess varying density (300 to 3000 kg/m^3 or more) and strength (5–6 to 120–160 MPa or more), and can be used as heat-insulating, protective or decorative and finishing articles, as also highly loaded supporting structures.

The important and most typical fields of practical application of polymer concretes have been quite clearly identified. These are: for decorative and finishing purposes, for producing sanitary and industrial equipment, for heat-insulating and chemically stable constructions, machine tool manufacture, engineering and electrical industries. Polymer concretes with special properties fall into a special group.

Depending on the field of application, specific requirements are prescribed for polymer concretes. For example, decorative and finishing materials (panels, window sills, flights of stairs, etc.) should possess highly decorative properties and should not be inferior in external form, hardness, and wear resistance to natural rocks (marble, onyx, high quality granites, etc.). These requirements are also applicable to polymer concretes which are used for producing sanitary and industrial equipment (basins, toilet bowls, bath tubs, cans, etc.).

The main indexes for heat-insulating polymer concretes are the density and thermal conductivity of the material. In works exposed to various aggressive media, the property of polymer concretes that is most important is their chemical stability.

Polymer concretes used for machine tool construction and in the engineering industries should possess high dimensional stability under the action of various external factors and high damping characteristics while those used in the electrical engineering industry should possess high dielectric properties. This group of polymer concretes includes concretes with high electrical conductivity, thermal stability, vacuum tightness, biological resistance, and high resistance to various types of radiations.

However, the short- and long-term strengths of these materials represent common and extremely important indexes of all types of polymer concretes.

4.1. Effect of Specimen Size and Loading Rate on Strength Characteristics of Polymer Concretes

It is known that polymer concretes, in terms of their strength properties and deformability, occupy an intermediate position between the polymer materials and ordinary concretes. It is this feature that calls for specific requirements for their test methods, evaluation of results, and ultimately for designing load-bearing structures. Thus, the presence of a coarse aggregate fraction in the polymer concrete predetermines the minimum specimen sizes, while the binder based on synthetic resins controls the loading rate of such specimens during the tests.

Polymer materials, unlike low-molecular-weight materials and depending on the rate of deformation, temperature, and stress state, can possess elastic, highly elastic, and irreversible plastic deformations. Highly elastic deformation, characteristic of polymer materials alone, is characterised by a low modulus of elasticity and large reversible deformations. Apart from the usual factors, temperature and duration for which the specimen is subjected to the influence of external forces exert a significant influence on the strength of polymers.

Therefore, for a correct evaluation of the strength characteristics by testing the polymer materials including polymer concretes, external stress σ, deformation ϵ, temperature t and test duration t should be taken into consideration. At a stable temperature and a constant loading rate, the short-term compressive strength characteristics of polymer concretes are determined mainly using the following equations: $\epsilon = f(\sigma)$ and $\sigma = f(\epsilon)$. Further, tests are carried out with a constant amplitude of stresses ($\sigma = $ constant) or with a constant amplitude of deformations ($\epsilon = $ constant). The dependences $\epsilon = f(t)$ and $\sigma = f(t)$ are determined while carrying out long-term tests.

For determining the effect of specimen sizes on compressive strength characteristics, a series of specimens was produced (Table 4.1). The statistical processing of the results of testing for axial compression of cubes of FAM and PN polymer concretes showed (Table 4.2) that the strength of polymer concrete specimens increases steadily with an increase in size from 30 mm $\times$ 30 mm $\times$ 30 mm to 150 mm $\times$ 150 mm $\times$ 150 mm unlike cement mortar specimens (Fig. 4.1). This feature of polymer concretes can be explained by their high strength characteristics and the more favourable test conditions.

According to A.E. Desov, the axial compressive strength of cement concrete for a given size of aggregates increases with the increasing (to a certain limit) size of the specimen when the surface friction between the two platens of the machine and the test specimen is eliminated. This is particularly strikingly seen in the case of high-strength cement concretes. From the physical point of view,

Table 4.1. Size of specimens

Size of specimens, mm	Aggregate size, mm	Area of cross-section of specimens, cm^2
	Cube specimens	
30 × 30 × 30	3–5	9
50 × 50 × 50	5–10	25
70 × 70 × 70	5–10	49
100 × 100 × 100	5–10	100
150 × 150 × 150	5–10	225
	Prism specimens	
40 × 40 × 160	3–5	16
50 × 50 × 150	5–10	25
70 × 70 × 210	5–10	49
100 × 100 × 400	20–25	100

Table 4.2. Results of tests of axial compression of specimens of FAM and PN polymer concretes

Size of specimens, mm	No. of specimens	Mean arithmetical strength, MPa	Mean square deviation	Average error of mean arithmetical deviation	Variation coefficient	Degree of uniformity
			Cube specimens			
30 × 30 × 30	10	70	12.3	3.84	1.76	0.48
50 × 50 × 50	12	78.3	10.5	3.18	1.34	0.62
70 × 70 × 70	12	84.4	10.4	2.97	1.23	0.63
100 × 100 × 100	12	92.7	8.6	2.6	0.92	0.73
150 × 150 × 150	9	96.1	8	3.2	0.87	0.75
			Prism specimens			
50 × 50 × 150	12	77.2	11.2	3.41	1.41	0.6
70 × 70 × 210	9	85.3	10.8	3.2	1.32	0.61
100 × 100 × 400	9	91.9	9.3	2.8	1.12	0.72

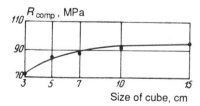

Fig. 4.1. Dependence of compressive strength of polymer concrete specimens based on FAM furan resin on the size of cube specimens.

this is explained by the fact that the number of fracture planes in the limiting state is proportional to the size of the aggregates.

It follows from Desov's hypothesis that the theoretical resistance to axial compression of specimens with identical cross-section is independent of the height of the specimen to a certain extent. Thus, the strength of cube and prism specimens of high-strength concretes should be identical.

The high strength of polymer concretes and the presence of a separating film on the specimen surface after coating the mould make for the required conditions for an experimental confirmation of Desov's hypothesis.

The strength of cube and prism specimens of polymer concretes varied in the scatter range of test data and almost wholly confirmed the data obtained for cement concrete specimens (Table 4.2).

For determining the effect of the loading rate on the strength characteristics, tests were carried out at 18 to 20°C and at a relative atmospheric humidity of about 60%. The test duration varied from 10 to 10,000 sec with an interval of one-tenth of the range. For each composition of the polymer concrete and the type of tests, the number of specimens was 30 to 40 for each type of stress state. In all, about 800 specimens were tested.

The test results point to the significant influence of the loading rate on the strength of polymer concretes tested in all types of stress states.

At low loading rates, the dependence of strength R on the rate of loading V_σ is manifest to the maximum extent. With an increasing loading rate, this dependence weakens (Fig. 4.2).

Based on the statistical processing of the results of testing polymer concrete specimens at different loading rates, the correlation factor for the coordinates $R-\lg t$ was determined. The correlation equation for strength reduction of FAM polymer concrete in static compression tests in the range 10 to 10,000 sec has the following form:

$$R = M_\sigma + r(m_\sigma/m_t)(\lg t - M_t) = 1056.3 - 59.1, \qquad \dots (4.1)$$

where M_σ is the mean arithmetical value of ultimate strength; r the correlation coefficient; m_σ the mean square deviation; and m_t the mean square deviation for the logarithm of test duration.

Assuming that the strength of the specimens at loading duration 10 sec is 100%, the dependence of strength on loading rate can easily be established (Table 4.3).

Figs. 4.3 and 4.4 depict correlation curves which strikingly characterise the time dependence of the strength of polymer concretes. Under all types of stress states, PN-1 polymer concretes exhibit deformability to a large extent and as a consequence the dependence $R = f(\lg t)$ for these polymers is more explicit compared to the FAM polymer concretes. Confidence ranges with 0.95 probability are marked in all the graphs.

These data show that the scatter of strength characteristics is minimal in the zone of weak dependence $R = f(V_\sigma)$ and depends mainly on the uniformity of the material.

62

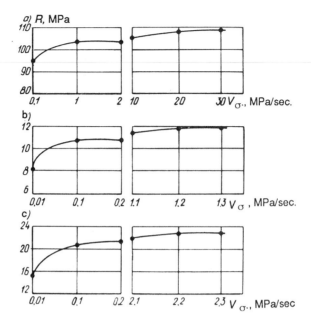

Fig. 4.2. Dependence of strength of polymer concrete specimens on loading rate in compression (a), tensile (b), and bending tension (c) tests.

Tests on a significant number of samples and the processing of test results by methods of mathematical statistics helped to draw the following conclusions:

— Variation coefficients of mechanical characteristics for polymer concretes based on PN-1 resin are less than for FAM polymer concretes (Table 4.3). This points out that polymer concretes based on PN-1 resin have a more uniform structure and high strength compared to the FAM polymer concretes;

— For testing polymer concrete specimens in all types of stress states, the recommended test duration varies from 100 to 150 sec, to which the following loading rates correspond: tensile and bending tests 0.1–0.2 and compression test 1–1.2 MPa per sec. A minor increase of the loading rate practically does not affect the test results; and

— The reduction of strength limits as the test duration changes from 10 to 10,000 sec is practically independent of the type of stress state and is 25–30% for polymer concretes based on polyester resins and 15–18% for those based on furan resins.

According to the method adopted, the prism compressive strength and modulus of elasticity of cement concretes is determined by successive loading of the prism in stages of roughly 0.1 each of the anticipated breaking load. The

Table 4.3. Dependence of strength of specimens on loading rate

Type of test	Mean arithmetical strength at $t = 10$ sec, MPa		Correlation factor for strength on loading duration (in the range 10 to 10,000 sec), %		Mean error of $m_{\sigma t}$, MPa	Coefficient of strength reduction on increasing the duration $K = \sigma_{10000}/\sigma_{10}$	
	FAM polymer concrete	PN polymer concrete	FAM polymer concrete	PN polymer concrete		FAM polymer concrete	PN polymer concrete
Compression	92.72	112.4	$R_{comp} = 100 - 5.59 \lg t$	$R_{comp} = 100 - 7.31 \lg t$	$53 - 3.6$	$81/99.72^{+} = 0.812$	$85.84/112.4 = 0.763$
Bending	15.13	22.14	$R_b = 100 - 5.2 \lg t$	$R_b = 100 - 8.91 \lg t$	$1.6 - 0.7$	$12.92/15.13 = 0.853$	$15.62/22.14 = 0.706$
Tensile	7.03	11.85	$R_t = 100 - 4.53 \lg t$	$R_t = 100 - 8.98 \lg t$	$0.9 - 1$	$6/7.03 = 0.853$	$8.4/11.95 = 0.704$

+ *Sic*: should read 81/92.72—Technical Editor.

64

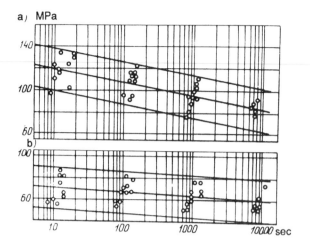

Fig. 4.3. Correlation dependence of compressive strength on loading rate for polymer concrete specimens based on PN-1 (a) and FAM (b) resins.

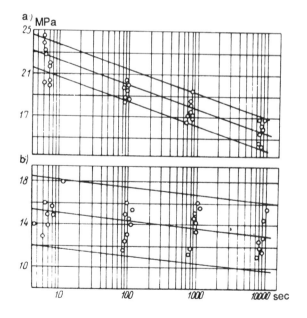

Fig. 4.4. Correlation dependence of bending tensile strength on loading rate of polymer concrete specimens based on PN-1 (a) and FAM (b) resins.

retention time at each stage of loading is 4–5 min followed by relieving and retention for 4–5 min and then loading in the next stage. Thus, according to this method, the total duration of determining the initial modulus of elasticity of each prism is roughly 60 min.

Considering the significant effect of loading rate on the strength properties of polymer concretes, we proposed a method for determining the modulus of elasticity by successive loading of the prism in stages of 0.1 each of the anticipated breaking load without relieving or retention at each stage. The total test duration in this case was roughly 5 min.

Tests were carried out on prisms of 70 mm × 70 mm × 280 mm and 100 mm × 100 mm × 400 mm. Strain gauges were connected to a device which automatically measured the deformations. Longitudinal gauges were glued to all the four faces of the prism and transverse gauges to the two opposite faces.

The test results (Table 4.4) point to a direct dependence between the strength of the polymer concrete and the initial modulus of elasticity and inverse relation between the initial modulus of elasticity and Poisson's ratio for each type of binder.

Table 4.4. Dependence of compressive strength of polymer concrete on modulus of elasticity

Polymer concrete	No. of samples tested	Prism compressive strength R_{pr}, MPa	Initial modulus of elasticity $E \times 10^3$, MPa	Poisson's ratio
FAM with andesite filler	12	90	26	0.22
FAM with graphite filler	12	72	23	0.246
PN with andesite filler	10	110	22	0.262
PN with graphite filler	10	95	20.5	0.273

4.2. Evaluation of Homogeneity of Polymer Concretes

A preliminary evaluation of the homogeneity of polymer concretes can be made by sounding the polymer concrete prisms in an ultrasonic device. The propagation duration of ultrasound through the specimen is related to the dynamic modulus of elasticity as follows:

$$E_d = (l^2 \rho / t^2) K, \qquad \ldots (4.2)$$

where E_d is the dynamic modulus of elasticity of the material, Pa; l the length of the specimen, cm; ρ the density of the specimen, g/cm^3; and K a coefficient depending on the ratio of the wavelength and specimen size, for the specimen adopted, $K = 1$.

The polymer concretes may be regarded as quite homogeneous when the dynamic moduli of elasticity of the tested specimens differ from each other by ±10–15%.

More reliable data on the homogeneity of polymer concretes are obtained by statistical processing of a large number of specimens produced simultaneously with the fabrication of a large series of structures under production conditions.

When fabricating 96 electrolytic baths using polymer concretes based on furan resins at a metallurgical combine for copper electrolysis, six cubes of 150 mm × 150 mm × 150 mm were produced simultaneously with the moulding of each bath. Three of these cubes were tested after hardening for 28 days at normal temperature and the rest after retaining for one day at normal temperature followed by heating at 80°C for another day. Nearly 600 cubes were thus produced and tested.

Statistical processing of the results of short-term tests of cubes hardened at normal temperature for 28 days and the cubes hardened by heating showed (Table 4.5) that polymer concretes in the first case have very low strength characteristics, positive asymmetry, and a significant negative excess. Further, the variation coefficients in both cases practically coincided while the accuracy indexes were quite proximate.

Table 4.5. Results of testing cube specimens of polymer concretes hardened by heating

Method of hardening polymer concrete specimens	Mean arithmetical strength, MPa	Mean square deviation	Variation coefficient	Accuracy index P, %	Asymmetry A	Excess E	A/m$_A$	E/m$_E$
Retention for one day at normal temperature and heating for another day at 80°C	71.1	±9.03	1.27	1.47	−3.34	−0.2	−11.733	−0.35
Retention for 28 days at normal temperature	69.9	±8.84	1.36	1.97	0.31	−1.05	0.843	−1.41

The following conclusions can be drawn by comparing the experimental and normal curves of the distribution of short-term strength of polymer concrete (Fig. 4.5) and also the results of statistical processing (Table 4.5): the ratio of asymmetry and excess indices to their errors in the first as well as in the second case is much less than three. Thus, with reasonable justification, it may be affirmed that the ultimate short-term compressive strength of the polymer concretes tested obeys the law of normal distribution. However, when hardening by heating, the experimental curve of distribution shows a significant negative asymmetry pointing to the predominance of variants with high values compared to the arithmetical mean value; in the second case, the asymmetry is positive and hence variants with low values predominate. In the first case, an insignificant negative excess $E = -0.2$ is noticed while in the second case $E = -1.05$, pointing to highly uneven hardening and hence significant inhomogeneity of the polymer concrete.

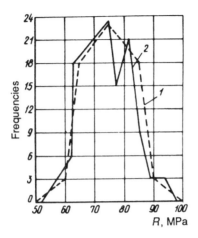

Fig. 4.5. Distribution curves of short-term strength of polymer concrete specimens based on FAM resin:

1—normal; 2—experimental.

Having established that the experimental distribution curves obey the normal law of distribution and by knowing the coefficient of variation, the coefficient of homogeneity of the polymer concrete can be determined as under:

$$K_{1-\text{hom}} = (10 - 3V)/10 = [10 - (3 \times 1.27)]/10 = 0.62, \quad \ldots (4.3)$$

$$K_{2-\text{hom}} = (10 - 3V)/10 = [10 - (3 \times 1.36)]/10 = 0.59. \quad \ldots (4.4)$$

Thus, statistical processing of the results of a large number of works' tests established quite reliably the possible limits of the coefficient of homogeneity of FAM polymer concretes and provided the basic theoretical characteristics.

Basic Theoretical Characteristics of FAM Polymer Concretes Obtained on the Basis of Short-term Physical and Mechanical Tests

Grade according to compressive strength $\overline{R}$, MPa . . . 70–90
Grade according to tensile strength $\overline{R}_{\text{st}}$, MPa 5–10
Theoretical modulus of elasticity E_0, MPa . . . . $(2.3–2.75) \times 10^4$
Theoretical short-term deformation on axial
   compression ϵ_s . . . . 2×10^{-3}
Theoretical short-term deformation on axial
   tension $\epsilon_{\text{s.t}}$ . . . . $(2–2.5) \times 10^{-4}$
Cohesion with steel reinforcement, MPa 6–8
Coefficient of thermal deformation, 1/°C . . . . $(12–15) \times 10^{-6}$

Cold resistance (number of cycles) . . . . Not less than 300
Coefficient of working conditions K_0 for theoretical
   compressive strength and theoretical modulus of
   elasticity at $°C$:
    20 1
    40 . . . . 0.9
    60 . . . . 0.8
    80 . . . . 0.7
Coefficient of homogeneity K_{hom} . . . . 0.6 to 0.65

Innumerable tests on different types of polymer concretes were carried out by the proposed method to determine the short-term strength characteristics. These were compared with the data of the statistical processing of tests carried out at several industrial establishments. An analysis of these data helped to obtain average physical and mechanical characteristics for different types of polymer concretes (Tables 4.6 to 4.12).

Table 4.6. Average physical and mechanical indexes for FAM (FA) polymer concretes

Index	Based on dense aggregates, composition 1	Based on porous aggregates, composition 2
Density, kg/m³	2200–2400	1500–1900
Short-term strength, MPa:		
under compression	70–90	30–60
under tension	5–8	3–5.5
Elastic modulus under compression, MPa	$(20–32)10^3$	$(13–20)10^3$
Poisson's ratio	0.2–0.24	0.19–0.21
Specific impact strength, J/cm²	0.15–0.25	0.1–0.2
Linear shrinkage during hardening, %	0.1	0.1–0.15
Water absorption in 24 h, %	0.05–0.3	0.1–0.4
Martens' thermal resistance, $°C$	120–140	120–140
Heat resistance, kJ/(m-h-deg)	$(0.65–0.75)4.18$	$(0.25–0.5)4.18$
Cold resistance not less than, cycles	300	300
Coefficient of thermal expansion, $1/°C$	12×10^6	$(12–13)10^6$
Abradability, g/cm²	0.018–0.21	0.025–0.35
Specific electric resistance:		
surface, ohm	3.7×10^{10}	3.7×10^{10}
volume, ohm-cm	3.8×10^3	5.8×10^3
Dielectric loss tangent at 50 Hz and 65%		
relative humidity	0.05–0.06	0.02–0.05
Inflammability	0.14	0.14

4.3. Long-term Strength of Polymer Concretes

When studying the long-term strength of polymer concretes, not only the strength characteristics but also their deformation kinetics under prolonged action

Table 4.7. Average physical and mechanical indexes for FAED polymer concretes

Index	Based on dense aggregates, composition 3	Based on porous aggregates, composition 4
Density, kg/m^3	2200–2400	1500–1800
Short-term strength, MPa:		
under compression	90–110	50–80
under tension	9–11	3–9
Elastic modulus under compression, MPa	$(32–38)10^3$	$(12–18)10^3$
Poisson's ratio	0.26–0.28	0.24–0.26
Specific impact strength, J/cm^2	0.35–0.45	0.2–0.3
Linear shrinkage during hardening, %	0.05–0.08	0.06–0.1
Water absorption in 24 h, %	0.01	0.2–0.5
Martens' thermal resistance, °C	120	120
Heat resistance, kJ/(m-h-deg)	(0.65–0.75)4.18	(0.25–0.5)4.18
Cold resistance not less than, cycles	500	300
Coefficient of thermal expansion, 1/°C	$(10–14)10^{-6}$	$(10–14)10^{-6}$
Abradability, g/cm^2	0.005–0.01	0.01–0.02
Dielectric loss tangent at 50 Hz and 65% relative humidity	0.04–0.05	0.03–0.05
Inflammability	1	1

of loads should be determined with adequate reliability.

The most widely used models of deformation process (Maxwell, Kelvin–Foigt, Alexandrov–Lazurkin, etc.) do not reflect quantitative dependences and are suitable only for a qualitative description of the process under stresses not exceeding the long-term strength of the material. These models, while mainly describing the initial and ultimate parameters, do not reflect the deformation kinetics or the nature of degradation in time.

It is well known that the long-term strength of polymer concretes is characterised by stresses at which the deformations of structures are of the damping type and can remain practically constant for a long time. Therefore, the complete information on the material behaviour under load should include not only the ultimate results, but also their kinetics.

The long-term strength of polymer concretes can be determined by processing experimental creep curves recorded when testing identical specimens loaded to different levels of stresses (Fig. 4.6). However, in this case, the long-term strength is determined only approximately since this method does not permit the accurate determination of the maximum level of stresses at which the deformations dampen in time.

The processing of a large number of test results of polymer concrete specimens and actual structures shows that the characteristic deformation curves depicted in Fig. 4.6 relative to the type of polymer concrete, extent of its reinforcement, and other factors may deviate from the mean values on either side

Table 4.8. Average physical and mechanical indexes for PN polymer concretes

Index	Based on dense aggregates, composition 5	Based on porous aggregates, composition 6
Density, kg/m^3	2200–2400	1500–1800
Short-term strength, MPa:		
under compression	80–100	50–80
under tension	7–9	2–8
Elastic modulus under compression, MPa	$(28–36)10^3$	$(12–18)10^3$
Poisson's ratio	—	0.8–0.22 [sic]
Specific impact strength, J/cm^2	0.2–0.25	0.1–0.2
Linear shrinkage during hardening, %	0.08–0.1	0.2–0.25
Water absorption in 24 h, %	0.05–0.1	0.05–0.3
Martens' thermal resistance, °C	80	80
Thermal conductivity, kJ/(m-h-deg)	$(0.6–0.7)4.18$	$(0.25–0.5)4.18$
Cold resistance not less than, cycles	300	300
Coefficient of thermal expansion, 1/°C	$(14–20)10^{-6}$	$(14–18)10^{-6}$
Abradability, g/cm^2	0.015–0.025	0.02–0.03
Dielectric loss tangent at 50 Hz and 65% relative humidity	0.03–0.06	0.01–0.04
Inflammability		
in PN-1 resin	2.1	2.1
in PN-63 resin	0.47	0.47

Table 4.9. Average physical and mechanical indexes for polymer concretes based on polyester acrylate resins

Index	MGF–9, composition 7	TGM–3, composition 8
Density, kg/m^3	2100–2300	2100–2300
Short-term strength, MPa:		
under compression	70–100	80–110
under tension	7–9	9–11
Elastic modulus under compression, MPa	$(14–15)10^3$	$(19–20)10^3$
Poisson's ratio	0.22–0.24	—
Specific impact strength, kJ/m^2	0.08–0.21	0.06–0.18
Linear shrinkage during hardening, %	0.09–0.12	0.11–0.15
Water absorption in 24 h, %	0.05–0.12	0.08–0.15
Thermal resistance, °C	105–110	120–128
Cold resistance not below, cycles	300	300
Coefficient of thermal expansion, 1/°C	—	$(14–16)10^{-6}$
Abradability, g/cm^2	0.007–0.018	0.006–0.015

but they objectively reflect the general patterns of the build-up of deformations under prolonged loading of polymer concretes and reinforced polymer concretes. These general patterns are strikingly visible when studying the structural dia-

Table 4.10. Average physical and mechanical indexes for KF–Zh polymer concretes

Index	Based on dense aggregates, composition 9	Based on porous aggregates, composition 10
Density, kg/m^3	2200–2400	1500–1800
Short-term strength, MPa:		
under compression	50–60	30–40
under tension	3–4	2.5–4
Elastic modulus under compression, MPa	$(10–14)10^3$	$(9–10)10^3$
Poisson's ratio	0.22–0.24	0.2–0.21
Specific impact strength, J/cm^2	0.15–0.25	0.1–0.2
Linear shrinkage during hardening, %	0.2–0.22	0.16–0.2
Water absorption in 24 h, %	0.1–0.3	0.2–0.6
Martens' thermal resistance, °C	120–100	100–120
Thermal conductivity, kJ/(m-h-deg)	(0.65–0.75)4.18	(0.4–0.5)4.18
Cold resistance not below, cycles	200	200
Coefficient of thermal expansion, 1/°C	$(15–16)10^{-6}$	$(13–15)10^{-6}$
Abradability, g/cm^2	0.02–0.03	—
Dielectric loss tangent at 50 Hz and 65% relative humidity	0.08–0.1	0.06–0.1
Inflammability	0.2	0.2

Table 4.11. Average physical and mechanical indexes for MMA polymer concretes

Index	Based on dense aggregates, composition 11	Based on porous aggregates, composition 12
Density, kg/m^3	2200–2400	1500–1800
Short-term strength, MPa:		
under compression	70–90	40–65
under tension	10–13	5–8
Elastic modulus under compression, MPa	$(10–15)10^3$	$(8–10)10^3$
Poisson's ratio	0.26–0.28	0.25–0.27
Linear shrinkage during hardening, %	0.15–0.2	0.2–0.25
Water absorption in 24 h, %	0.01	0.05–0.2
Martens' thermal resistance, °C	60	60
Thermal conductivity, kJ/(m-h-deg)	(0.65–0.75)4.18	(0.25–0.5)4.18
Cold resistance not less than, cycles	500	300
Coefficient of thermal expansion, 1/°C	$(12–16)10^{-6}$	$(12–18)10^{-6}$
Dielectric loss tangent at 50 Hz and 65% relative humidity	0.04–0.05	0.02–0.04
Inflammability	2.1	2.1

grams proposed by A.I. Chebanenko [19] which describe the limiting state of the material under different types of static loading (Fig. 4.7).

72

Table 4.12. Average physical and mechanical indexes for polymer silicate concretes

Index	[Based on dense aggregates]* Composition 13	[Based on porous aggregates]* Composition 14
Density, kg/m³	2100–2300	2300–2500
Short-term strength, MPa:		
under compression of cubes	28–30	25–35
under compression of prisms	21–23	20–25
under tension	2.5–3	3.5–4
Elastic modulus, MPa	25,000	25,000
Poisson's ratio	0.22	0.2
Specific impact strength, J/cm²	0.16	0.15
Linear shrinkage during hardening, %	0.15	0.15
Water absorption in 24 h, %	6	8
Martens' thermal resistance, °C	300–350	300–350
Thermal conductivity, kJ/(m-h-deg)	0.5×4.18	0.5×4.18
Cold resistance, cycles	80	80
Coefficient of thermal expansion, 1/°C	8×10^{-6}	8×10^{-6}

*Omitted in Russian original—Translator.

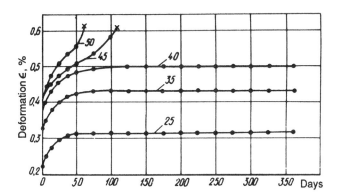

Fig. 4.6. Creep curves under compression of prismatic specimens of FAM polymer concretes.

Values on curves represent the load, MPa.

The boundary line of this diagram OK (K_t) characterises the interdependence between the stresses σ and deformations ϵ on rapid and continuous static loading (growth rate of stresses should be not less than 60 MPa/min). At some level, depending on the position of point K, before the strength is drained the proportionality between σ and ϵ breaks down due to the generation and growth of microfissures of elastic particles in the material.

The position of point K is determined by the state at which perceptible

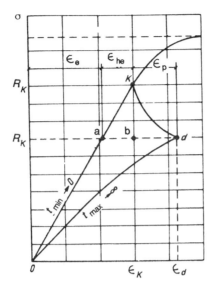

Fig. 4.7. Diagram of limiting states for polymer concretes.

residual deformations arise as fixed under conditions of rapid relaxation of the specimen. If the static loading and relaxation are effected at stress intensities not exceeding a certain level, i.e., at $\sigma < R_K$, residual deformations do not arise and, conversely, at $\sigma > R_K$, these do arise and, on subsequent repeated loading, advance right up to total disintegration of the material as a result of the build-up of microfissures.

Under conditions of prolonged loading $(t_{max} \rightarrow \infty)$ at fixed stresses, whose magnitude does not exceed the ultimate long-term strength of the material R_d, the deformation of the specimen is characterised by boundary curve Od (Fig. 4.7); each point on it determines the equilibrium state of deformations and stresses. In this case, the limiting equilibrium deformation of polymer concrete ϵD consists of three sections:

$$\epsilon_d = \epsilon_e + \epsilon_{he} + \epsilon_p,$$

where ϵ_e is the elastic reversible deformation characterised by point a on line OK at $\sigma = R_d$; ϵ_{he} the maximum high-elastic deformation which is reversible in time; and ϵ_p the maximum plastic deformation.

The high-elastic and plastic deformations are manifest in time and characterise the creep of the polymer concrete. The section $ad = \epsilon_{he} + \epsilon_p$ corresponds to the full section of creep under conditions of fixed stress intensity equal to the long-term strength of the material R_d.

Now, if the transformation process of point K into point d on steady reduction of stresses is depicted such that the stress deformed state of the material falls on the boundary of the commencement of microfissures of elastic particles, we obtain a curve like Kd. The process of transformation of point K is accompanied by the relaxation of stresses and the formation of high-elastic and plastic deformations of the materials until an equilibrium state is reached at point d.

The increment of deformations characterised by zone Kbd and enclosed by curve Kd arises due to viscous flow while the rest of the creep is determined by zone Kba of the high-elastic phase adjoining boundary line OK.

Within the diagram $OKdO$, all the intermediate values of polymer concrete deformations produced at different intermediate growth rates of stresses can be established. The enclosing curve of the limiting state Kd fixes the moment of the onset of intense microfissuring process of the elastic phase which is followed by the disintegration of the material.

However, the above discussion applies to the use of polymer concretes under relatively stable temperature conditions. On significant temperature changes, appropriate corrections will have to be made in the evaluation of the temperature—time dependence of the properties of polymer concretes.

A method which takes into consideration the temperature effects has been proposed in [6]. According to the authors of this method, since the long-term strength of polymer materials is determined by the entire history of the effect of stresses and temperatures and the temperature of vitrification formation of the polymer, in turn depends on the stress applied, the limiting zones of stresses, temperatures, and time must be indicated for a correct and complete evaluation of the efficiency of such materials. These boundary regions are established with the help of three-dimensional diagrams in co-ordinates σ, t, and $\ln t$ (Fig. 4.8). In this case, the surface corresponding to strength is described by the equation

$$\ln R_d = \ln t_0 + (V_0 - \gamma\sigma)/KT, \qquad \ldots (4.5)$$

where R_d is the long-term strength; t_0, V_0, and γ the constants of the material; T the absolute temperature; and K the universal gas constant.

The surface corresponding to stress relaxation is described by the equation

$$\ln t_r = \ln t_{or} + (V_{or} - \gamma_r\sigma)/KT, \qquad \ldots (4.6)$$

where t_r is the duration of relaxation; and t_{or}, V_{or}, and γ_r the relaxation constants.

Having determined experimentally the constants figuring in equations (4.5) and (4.6), the zone of efficiency of the polymer concrete can be determined by this method.

An analysis of the results of the industrial use of polymer concrete structures shows that they are most frequently used at temperatures of zero to 80–100°C. In this temperature range, the interrelation between the strength of the material and temperature is represented by a set of inclined lines converging at point B (Fig. 4.9). The upper straight line Bl_0 fixes the moment when the strength of the material is drained $(t_{min} \to 0)$ and corresponds to point l on the diagram (see Fig. 4.7). Line BK_0 (Fig. 4.9) determines the short-term static strength R_k

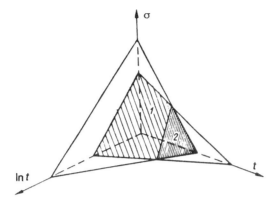

Fig. 4.8. Temperature—time characteristics of strength: 1—surface corresponding to strength; 2—surface corresponding to stress relaxation.

in the same loading regime and corresponds to the states fixed by points K (see Fig. 4.7). The lower inclined line Bd_0 determines the long-term strength of material Rd_0 at $t_{max} \rightarrow \infty$ and corresponds to point d (see Fig. 4.7). At 100°C, the corresponding minimum values of R_K and R_d are established.

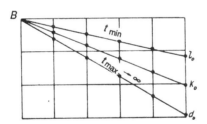

Fig. 4.9. Effect of temperature (t) and duration of loading (t) on the strength of FAM polymer concretes.

Using the technique developed by A.I. Chebanenko for plotting the diagrams, a method is proposed for constructing the diagrams of the limiting states in coordinates η–ϵ [62].

By introducing the parameter η_τ characterising the ratio of stresses fixed by congruent curves $\eta_k \eta_d$ to the corresponding stress σ_K determined by boundary line OK (Fig. 4.10), this parameter is found to fall in the following range near the enclosing curve of limiting states:

$$\eta_d = R_d/R_K \leq \eta_\tau = R_\tau/R_K \leq 1 \ . \qquad \ldots (4.7)$$

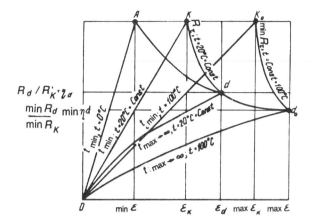

Fig. 4.10. Diagram of the limiting states of FAM polymer concrete at zero to 100°C.

Evidently, the value $\eta_\tau = 1$ corresponds to boundary line OK of the loading sector (Fig. 4.7) including the limiting state fixed by point K. The line of equilibrium deformations Od and its extreme point d correspond to the minimum values of parameter $\eta_\tau = \eta_d$. In principle, parameter η_τ characterises the loading regime of material, since σ_τ and R_τ are directly related to the growth rate of loading [27]. Curves Kd fall in the zone of limiting states AK_0d_0A (Fig. 4.10).

At zero temperature, curve Kd is transformed into point A and sector OKd into line OA. At 100°C, boundary curve K_0d_0 is formed and encloses the zone of limiting states and loading sector OK_0d_0. The lower boundary of the zone of limiting states, line Add_0, fixes the value of parameter η_d and the corresponding deformations ϵ_d which does not evidently run beyond the following limits:

$$\min R_d / \min R_K = \min \eta_d \leq \eta_d = R_d/R_K \leq 1, \qquad \ldots (4.8)$$

$$\max \epsilon_d \geq \epsilon_d \geq \min \epsilon. \qquad \ldots (4.9)$$

The limits of deformations ϵ_k corresponding to short-term strength R_k may be seen in Fig. 4.10:

$$\min \epsilon \leq \epsilon_k \leq \max \epsilon_K. \qquad \ldots (4.10)$$

In this case, any intermediate sector of loading and curve Kd enclosing it correspond to a definite temperature: T = const t.

The mathematical interpretation of the boundary lines of diagram AK_0d_0A in the co-ordinates of marginal points A, K_0, d_0, and intermediate points d and K poses no difficulty. The mechanical characteristics corresponding to these points are established experimentally.

For a given polymer concrete mix, the zone of limiting states AK_0d_0A may serve as a typical index of the mechanical properties of structural polymer concrete within the temperature zone zero to 100°C. This zone can be extended when so required, however, to the actually required values of negative temperatures.

Thus, an analysis of the structure of the diagram of limiting states at normal temperature and in coordinates $\eta–\epsilon$ helps to express the temperature—time dependence of long-term strength and deformability of polymer concretes with accuracy adequate for practical purposes, not in a three-dimensional [27] but two-dimensional plane diagram. Further, the load-bearing capacity of polymer concrete can be evaluated not only from the first limit state, i.e., strength, but also from the second limit state, i.e., deformability. In many cases, when considerable statistically processed test results of a given type of polymer concrete are available, the load-bearing capacity is calculated by a simpler method, taking into consideration the effect of temperature and other external influences and by using the corresponding reducing coefficients (see Chapter 8).

The experimental-theoretical method of evaluating the load-bearing capacity of polymer concretes and the processing of the results of long-term tests of specimens and actual structures helped in deriving the theoretical characteristics of polymer concretes under conditions of prolonged loading and a suitable method for designing such structures (see Chapter 8).

4.4. Polyester Resins used for Producing Decorative and Finishing Products (Federal Republic of Germany)

From the earliest times, marble was used for decorative finishing of the most impressive interiors of buildings. The collective term marble represents granular—crystalline limestones formed from ordinary limestone. Marble possesses diverse properties depending on its genesis, location, and age. Its compressive strength is 80–200 N/mm^2, abrasion resistance 5–25 cm^3/50 cm^2, and water absorption 0.2–0.6% by weight.[1]

These fairly scattered data show how very important it is to take into consideration the zone of application and the material reserves. The data given pertain in fact only to 'healthy marble' without fissures, cracks, stratified sections, and mined from blocks not subjected to erosion and not containing extraneous inclusions. Marble suffers from intense damage during transport and processing which ultimately affects its value. Moreover, the blocks from which panels are cut are of limited sizes.

The material in the form of window sills, table-tops, staircase or floor coverings is exposed to the action of water flowing from flower pots (containing for example, mineral fertilisers), spilled drinks, contaminated water, washing agents, abrasion, atmospheric action, etc. Such effects usually leave their mark on the surface and hence attempts are made to protect it by a layer of unsaturated transparent polyester resin.

---

[1] The units of measuring the physical values were worked out by the author of this section, R. Kreis.

While carrying out studies aimed at ensuring the development and supply of a new material not containing the defects of natural stone (without breakages in particular) but possessing similar properties without being inferior to the natural material, we took advantage of the advancements in chemistry as also in concrete production technology. Mixtures of mineral, granulated and finely ground aggregates of different size fractions and fineness of grind were formulated; the materials contained pigments bound with synthetic resins, hardeners, and promoters, i.e., these represented polymer slurries.

Unsaturated polyester and other resins may be regarded as the most preferred synthetic binding agents. Aggregates based on the size composition curve are represented by silica sands and powders of extremely high quality and purity used after heat drying.

The colouring agents for the slurries belong exclusively to Class 8 colour stability in the West German Wollaston scale. The idea is to use the most stable colouring material produced.[2] The panels are coloured throughout the thickness of the material and the marble finish is uniform.

In the course of special investigations, for example, for window sills, the following optimum technical data were prescribed; compressive strength $\approx$ 120 N/mm^2, tensile bending strength $\approx$ 25 N/mm^2, thermal stability from -40 to $+80°$C, thermal conductivity 0.8×1.163 W/(m-$°$C), water absorption less than 0.1% by weight, and abrasion resistance 4.9 cm^3/50 cm^2. A slurry of uniform composition ensures these parameters in the finished products.

When producing these products, surfaces totally free from pores, of uniform colouration, and without fissures and stratified sections in the internal structure are ensured.

For example, panels are produced (conforming to the standard) in lengths of up to 3500 mm and widths up to 600 mm and thickness of 20 mm. Panels of these sizes help the supplier to meet quickly all the requirements of the consumer for making, say, window sills.

There is no danger of chipping or peeling of the material when cutting the panels. There are no fissures or breakages. The surface of the cut edge is smooth and cutting can be done using the simplest of technical instruments used for sawing natural building stones. The consumer can obtain panels cast accurately to the required dimensions. In such a case, the fabricator need not cut the panels to the required size.

In particular, because of the high tensile bending strength in the range 25 N/mm^2, panels of length 3500 mm, width 600 mm, and thickness 20 mm can be produced. This ensures the absence of joints and makes it possible to support the panels at both ends without fear of breakage. In such cases, panels of similar sizes made from natural material or from non-reinforced concrete

---

[2] It may be noted that the requirement of one of the researches of the International Cement Association has been satisfied in this manner and that pigmented cement has a great future.

inevitably break. Using less mechanised means, the finishing personnel can cut or join the panels, drill holes in them, mount them on openings, etc.

Chemical Stability of Polymer Concretes Based on
Standard Polyester Resins Produced by Companies in FRG

Acetone	−
Ethyl alcohol 96%	−
Battery acid	+
Formic acid	+
Concentrated ammonia	−
5% ammonia	−
Gasoline (standard and super)	+
Benzene	−
10% calcium chloride solution	+
Solar oil (diesel fuel)	+
Concentrated acetic acid	−
10% acetic acid	+
30% solution of formaldehyde	+
Glycol	+
Petroleum residue	+
Isopropyl alcohol	+
Common salt solution (any concentration)	+
Machine oil	+
Methyl alcohol	−
10% lactic acid	+
12% sodium hypochlorite solution	−
20% caustic soda solution	−
5% caustic soda solution	−
85% phosphoric acid	+
10% phosphoric acid	+
Concentrated nitric acid	−
10% nitric acid	+
Concentrated hydrochloric acid	+
10% hydrochloric acid	+
Concentrated sulphuric acid	−
37.5% sulphuric acid (battery acid)	+
10% sulphuric acid	+
10% soda solution	+
Carbon tetrachloride	+
Toluene	+
1,2,2-trifluorotrichloroethane	+

Note: The use of polymer concrete products based on standard polyester resins is permissible (+) and not permissible (−) in the aggressive media indicated.

It is known that polymer concretes based on polyester, epoxide, furan, and other resins on prolonged retention in various aggressive media generally lose 40–50% of their strength. Table 4.13 gives interesting data on the strength variation of polymer concretes based on vinyl ester resin relative to the retention duration in different media. The strength of this polymer concrete not only does not decrease, but in many cases records a significant increase.

Table 4.13. Change of compressive strength of vinyl ester concrete after storage in different media, N/mm²

Chemical reagent	Storage duration, days			
	0	7	30	120
Distilled water	112	116	122	122
HCl, 10%	112	119	123	121
H_2SO_4, 20%	112	117	118	127
Acetic acid, 25%	112	114	116	116
Acetic acid, 50%	112	117	114	120
HNO_3, 5%	112	112	113	114
HNO_3, 15%	112	112	111	120
NaOH, 10%	112	113	111	114
NaOH, 20%	112	113	111	116
Toluene	112	115	122	131

For protecting concrete products and structures, various polymer coatings are usually found to be extremely effective.

In regions with extreme temperature variations of the external aqueous environment, water penetrates the concrete, freezes, expands and generates stresses giving rise to fissures and cracks in the concrete. The transport of conventional precast concrete structures by heavy transport media and also the loads acting on them during transport cause intense wear of the surfaces of concrete structures giving rise to unfortunate accidents.

To prevent the wear of concrete structures, to protect the concrete against the penetration of water and chloride compounds, and also to restore shear strength (slip resistance) of fairly worn out surfaces, thin polymer concrete coatings can be applied. These coatings provide a barrier against the penetration of water, salts of thawed waters, and also other chemical compounds, and simultaneously ensure the stability of the structures for a long period.

Some of the coatings were produced for several years using binders of various types were very good, but the rest proved less effective. It was also established that a given protective system behaves differently under different conditions of application. This may be explained as due to sensitivity to moisture, chemical incompatibility, shrinkage, high degree of wear, poor quality of preparation of the undersurfaces and poor quality of workmanship in applying the protective coatings.

At present, a large number of factors have been established as affecting these materials and it is hardly possible to resolve the question of their protection by applying just a single coating.

An example of the effective application of a coating containing a polymer solution is the Brooklyn bridge in New York.

Tests on the protective coating method have been underway for several years and have shown that the protective material should possess the following properties: modulus of elasticity less than 1200 N/mm^2, ultimate elongation not less than 30%, tensile strength about 17 N/mm^2 and compressive strength about 40–60 N/mm^2.

A low modulus E is necessary to compensate for the usually high deviations between the indexes of thermal expansion of the polymer concrete and concrete based on cement binders. A relative ultimate elongation of the resin at not less than 30% should prevent internal stresses (or compensate for them) arising during temperature changes. Moreover, systems based on resin binders should be stable relative to the chemically aggressive alkaline base of concrete.

In the work on the Brooklyn bridge, good results were obtained by applying an epoxide system of a special composition on the entire surface. The same material was subsequently on the surfaces of other bridges in New York (Queensborough, Manhattan, and Williamsburg) and proved quite satisfactory.

To select the optimum materials, electrical resistance tests were carried out. This value should exceed 500,000 ohms for the material to remain impermeable to water and salts of thawed waters. Tests on shear strength were carried out according to the technical specifications of ATM E-303-74 (Table 4.14).

Table 4.14. Results of testing specimens for electrical resistance

Type of coating	Resistance tested, ohms	Shear strength (according to ATM E-303-74)
Latex	66,750	80
Methacrylate solution 1	336,500	63
Methacrylate solution 2	2,342,000	65
Epoxide composition	9,050,000	82

A visual inspection of the tested portions revealed wear and small fissures on the latex as also on both types of methacrylate coatings. Table 4.15 shows the mean physical and mechanical characteristics of different types of polymer concretes based on resins according to the data of Prof. E. Ohama (Japan).

Table 4.15. Physical and mechanical properties of typical polymer concretes

Index	Polymer concrete based on resins			Polymer concrete based on resins			Other materials	
	Furan	Polyester	Epoxide	Polyurethane	Phenol	Polymeth-acrylate	Asphalt concrete	Concrete with cement binder
Density, kg/m^3	2200–2400	2200–2400	2100–2300	2000–2100	2200–2400	2200–2400	2100–2400	2300–2400
Strength, MPa:								
compressive	70–80	80–150	80–120	65–72	50–60	80–150	2–15	10–60
tensile	5–8	9–14	10–11	8–9	3–5	7–10	0.2–1	1–5
bending	20–25	14–35	17–31	20–23	15–20	15–22	2–15	1–7
Elastic modulus, MPa $\times 10^3$	20–30	15–35	15–35	10–20	10–20	15–35	1–5	20–40
Water absorption, % by weight	0.05–0.3	0.05–0.2	0.05–0.3	0.3–1	0.1–0.3	0.05–0.6	1–3	4–6

5

Resistance of Polymer Concretes to Aggressive Media

5.1. Effect of Aggressive Media on Building Structures

In the aggregate, aggressive media can be gaseous, liquid, or solid and, in many cases, even multiphased.

The aggressive action of gaseous media on building structures is controlled by their nature, concentration and relative atmospheric humidity. The presence of gas liberations that are aggressive to building structures is characteristic of many industries. These include non-ferrous metallurgical, basic chemical, coking, petrochemical, synthetic fibre and many other industries.

Oxides of nitrogen and chlorine, hydrogen chloride, hydrogen fluoride, sulphur dioxide and hydrogen sulphide are most widely distributed and at the same time the most aggressive.

All gases, with the exception of ammonia and oxygen, represent acidic or acid-forming substances. Acid formation arises from them only in the presence of moisture in the form of droplets in the atmosphere or on the surface of the structures (mist or condensate). Therefore, a high atmospheric moisture is regarded as a factor intensifying the corrosion process on the surface layers of building structures [7].

Three levels of moisture saturation are quite distinctly distinguished depending on the 'moisture threshold': the zone of dry gases at atmospheric humidity up to 60%; moisture threshold at atmospheric humidity from 60 to 75%; and the zone of wet gases at humidity exceeding 75% (Fig. 5.1).

The first stage of saturation represents a comparatively low atmospheric moisture content, i.e., up to 60% relative humidity, at which acidic gases practically do not damage cement concretes and in some cases even compact these concretes. The compacting action of carbon dioxide on cement concrete or so-called carbonisation is well known. Gaseous silicon tetrafluoride exerts an even more favourable influence on cement concrete.

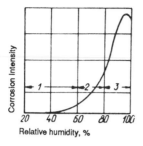

Fig. 5.1. Increasing aggressiveness of gases with increasing moisture content:

1—zone of dry gases; 2—moisture threshold; and 3—zone of moist gases.

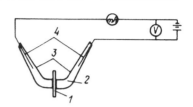

Fig. 5.2. Sketch showing device for determining diffusion coefficient:

1—specimen; 2—electrolyte; 3—glass adapters; and 4—platinum electrodes.

The second stage of high moisture saturation (60–75%) is already aggressive while the third stage with a moisture content of 75–95% is the most aggressive.

The critical threshold of moisture saturation decreases by 10–12% when hygroscopic dust particles are present in the atmosphere. By adsorbing moisture from the atmosphere and settling on the surface of the structure, such dust forms a high-moisture zone on the surface. This process is particularly characteristic, for example, of shops producing magnesium in the presence of carnallite dust.

In a building with 40–50% relative humidity, the walls and the ceiling can be wet. A similar local moisture prevails under the influence of hydrogen chloride vapours on concrete or plaster. The hygroscopic calcium chloride formed on the surface avidly absorbs moisture from the air.

The action of gas on porous structural materials (concrete, bricks, wood, etc.) is noticed not only on the surface, but also in very deep layers. According to the data available, gas penetration into low-density concrete can go up to 10 cm and in dense concrete up to 1–2 cm.

The aggressive properties of water are controlled by the extent of its mineralisation as also its acidity or alkalinity. Usually, river and lake waters record a weakly alkaline reaction. The total content of salts in river waters generally does not exceed 300–500 ml/l [sic].

Ground and subsurface waters usually contain mineral salts and other impurities. Marine (oceanic) water may contain up to 3500 ml [sic] salts/litre made up of sodium chloride 78%, magnesium chloride 11%, and magnesium, potassium, and calcium sulphates 4.7, 3.6, and 2.5% respectively.

Industrial effluents may contain extremely diverse admixtures, including solutions of salts, acids, and alkalis.

Even totally pure demineralised water may be aggressive towards many structural materials causing the leaching of lime and other dissolved constituents from cement concretes or benzenesulphonic acid from FAM polymer concretes. Moreover, water entering the micropores of the material causes an adsorptive reduction of strength.

Acids represent the most common and the most aggressive medium for several structural materials (cement concretes, silicate bricks, sedimentary rocks such as limestone, dolomite, etc.). Ceramic materials and concretes based on liquid glass are stable in acids but disintegrate relatively quickly in alkalis.

The aggressiveness of the acids is determined by their nature, concentration, pH of aqueous solutions, oxidising properties, and the temperature of the medium. The destructive action of acids and acidic gases depends also on the solubility of the corrosion products formed during their reaction with metals or concretes.

Mineral acids are generally more corrosive than organic acids. Among the latter, acetic, lactic and butyric acids are more aggressive. A particularly intense damage is noticed under alternating action of mixtures of different acids which is characteristic of non-ferrous metallurgy.

Oxidising agents exert maximum destructive action on many metallic structures, concretes and organic materials. Silicate materials alone remain unaffected under the action of oxidising media. Contrarily, fluorine-containing acids damage silicate materials. In these acids, graphite materials and polymer concretes based on furan resins with graphite fillers and aggregates are not affected.

Oxidation may be caused not only by the oxygen present in the air, but also in the acidic, neutral or alkaline media, containing some oxidising agents (Table 5.1).

Table 5.1. Important oxidising agents

Medium	Oxidising agent
Gases and vapours (with weakly acidic or neutral reaction in aqueous solution)	Halogens: chlorine, vapours of bromine and iodine, oxygen, and nitrogen (as constituents of air)
Intensely acidic	Acids: nitric HNO_3, chromic $H_2Cr_2O_4$, sulphuric H_2SO_4 (more than 70% concentration), persulphuric acid $H_2S_8O_8$.
Weakly acidic	Nitrous acid HNO_2 Hypochlorous acid $HOCl$ Bichromates $H_2Cr_2O_4 \cdot 2H_2O$; KCl_2O_7 Persulphates $H_2S_8O_8$ Hydrogen peroxide H_2O_2
Alkaline	Hypochlorites of sodium, potassium, and calcium ($NaOCl$, $KOCl$, and $CaOCl$)

Oxygen, invariably present in air, is the most widespread among the gaseous oxidising agents. The chlorine contained in the atmosphere of some industrial works is considerably more oxidizing than oxygen but its distribution in nature and the permissible concentration in the atmosphere of industrial works is hundreds of thousands of times less than oxygen.

Among the liquid oxidising agents, nitric and concentrated (over 70%) sulphuric acid, hydrogen peroxide and alkaline hypochlorites, which are capable of easily separating atomic oxygen, are of utmost importance.

Concentrated solutions of alkalis, especially at high temperatures, exert destructive action on many metals, rock materials and concretes.

The action of various salts on metals, concretes, ceramics, and polymer materials is less aggressive than acids or alkalis. The destructive action of salt solutions in many cases is dependent on their ability to react with water to form hydrogen (acidic) or hydroxyl (alkyl) ions; subsequent breakdown of the material proceeds in the same manner as under the action of acids or alkalis.

Many organic products of meat and milk, canning, distillery, and other branches of the food industry can be regarded as aggressive to cement concretes and other building materials. For example, solutions of sugar, fruit juices and syrups, on reacting with the lime present in cement stone, form soluble calcium saccharate. Vegetable and animal oils and fats represent complex esters of glycerine and fatty acids which, under the action of air and moisture, are oxidised and decomposed into glycerine and fatty acids [22, 24].

Degradation under the action of various radiations (gamma radiation, neutron radiation, etc.) represents a special type of destruction of building materials and structures. However, effects like those of various bacteria and other plant and animal organisms can be regarded as special types of effects.

From this short review, which is far from complete, it may be seen how diverse and various are the aggressive products and their action on building structures. It has been calculated that over 100 million tonnes of steel a year is lost all over the world as a result of corrosion. For example in the USA, according to the information of the American National Bureau of Standards, 40 million tonnes of steel is consumed in replacing corroded equipment alone [1]. At the scientific-technical conference of the Member Countries of the Council of Mutual Economic Assistance held in 1971 in Moscow to discuss the problems of developing protective measures against corrosion, it was pointed out that about 10% of industrial metal is lost due to various types of corrosion [1]. Investigations established that 15 to 75% of buildings and installations are exposed to the destructive action of atmospheric and industrial aggressive media.

At present, the number of chemical, synthetic fibre, non-ferrous metallurgical, cellulose-paper, printing, food, and many other industries has been steadily increasing all over the world. These use extremely diverse aggressive products, such as organic and inorganic acids, solvents and alkalis. The action of aggressive products in many cases may be combined with high pressure or vacuum, high voltage currents, and high or low temperatures. Thus, copper electrolysis is carried out in a 20% solution of sulphuric acid at 65–70°C and current 250 A/m^2. Some equipment in the hydrolysis industry operate at 80 to 175°C and pressures up to 0.8 MPa followed by rarefaction up to 0.05 MPa. The hydrolyzate contains sulphuric acid and mixtures of organic acids, i.e., formic and acetic. The production of plastics, synthetic fibres and many other materials is associated with the use of various solvents, acids and alkalis.

An inspection of several industrial works showed that not only apparatus and equipment but also building structures, i.e., bearing columns, collar beams, trusses, roofing sheets and panels, foundation blocks, floorings, and other parts of the buildings are also subjected to corrosion damage. In some cases, expenses on repair and restoration of industrial buildings and equipment amount to as much as the entire cost of the structure in the course of 4 or 5 years.

The ever-increasing annual investments in developing the chemical industry, non-ferrous metallurgy, food industries, and other branches of the national economy associated with using aggressive products give rise to continuously increasing expenses in fighting against corrosion, repair and restoration. Further, it should be borne in mind that the national losses due to corrosion are not restricted only to expenses in repair and restoration work; losses associated with work stoppages and the reduction of output as a consequence of these stoppages should also be attributed to corrosion.

As will be shown below, the application of chemically stable polymer concretes and reinforced polymer concrete structures under conditions of exposure to highly aggressive media is one of the most effective methods of resolving this complex practical problem.

5.2. Diffusion of Aggressive Liquids

The contemporary theory of the penetrability of polymers is based on the concepts of their structure and physical and mechanical properties. Kinetic and thermodynamic methods have helped in a rational understanding of the principle of diffusion of various liquids in high-molecular compounds. As a result of diffusion, the molecules of low-molecular-weight liquids penetrate the body of the material between the bonds of polymer molecules, fill the free spaces, shift the bonds and increase the distance between them; later, they separate the molecules and supramolecular aggregates of the polymer. As a result, the volume of the swollen polymer and its weight increase. The swelling process ceases after the intermolecular space of the polymer is completely filled by the low-molecular liquid.

Since the liquid diffuses at a low rate and is distributed unevenly throughout the thickness of the material, stresses arise in it even when deformation is not restricted to rigid external bonds. The outer swollen layers of the material tend to expand and draw toward themselves the internal 'dry' layers which resist this tension and restrict the tensile deformation. Therefore, along the section perpendicular to the diffusing liquid front, moist stresses of uneven value and sign arise in the material. The swollen layers are compressed while the internal layers are under tension. Naturally, the distribution patterns of moist stresses vary in time depending on the depth of penetration of the low-molecular material and as a result of relaxation processes. Moreover, as a result of the plasticising action of the liquid, the modulus of elasticity of the moist material will also change.

Microscopic, radiation, sorption, amperometric, and other methods are used to determine the diffusion coefficients of aggressive liquids in polymer materials.

The widely used sorption method essentially involves recording of the rate of liquid absorption by flat polymer specimens placed in an aggressive medium. On submersion, due to molecular thermal movement, an amount of the material Q is transported in time t through area S of the specimen perpendicular to the section of flow from a point with high concentration of the liquid to a point with low concentration. The aggressive liquid continues to move until a mobile equilibrium is established in the specimen.

The diffusion coefficient for polymer concretes is generally determined by the sorption method on disk-shaped specimens of diameter 160 mm and thickness 16 mm. This method, while being comparatively more accurate, is quite cumbersome and calls for prolonged tests over several months.

A method has been proposed [72] to determine the diffusion coefficient based on measuring the electrical conductivity of the specimens. The principle of this method lies in that, with increasing saturation of the specimen with the diffusing liquid, the body electrical resistance of the specimen falls. By using a graph showing the variation between current and time $I = f(t)$, the diffusion coefficient can be calculated. This method greatly reduces the test duration because of the activating action of the electric current.

The device for such tests (Fig. 5.2) is in the form of two adapters with platinum electrodes filled with an identical amount of electrolyte. The test specimen is fixed between the adapters using Mendeleev plaster. After connecting the circuit, current I is registered by a milliammeter, while voltage (V) at the terminals of the current source is measured by a voltmeter.

The electrical resistance of the specimen, R, is determined from the equation

$$R = r(\delta/S), \qquad \ldots (5.1)$$

where r is the specific resistance of the specimen; δ the thickness of the specimen; and S the cross-section of the specimen.

Hence,

$$r = RS/\delta. \qquad \ldots (5.2)$$

If it is assumed that the specific resistance of the specimen material is inversely proportional to the amount of the solution absorbed,

$$r = 1/QY, \qquad \ldots (5.3)$$

where Y is the equivalent electric conductivity, then

$$Q = \delta/RSY, \qquad \ldots(5.4)$$

where Q is the weight gain of the specimen in a state of equilibrium.

Since the solution in the adapters is an electrolyte, an identical amount of opposite charges or charged ions flows toward the specimen from both sides on passing an electric current. The concentration of the electrolyte in the specimen remains constant in a state of equilibrium. Therefore, the equivalent electrical conductivity Y will also be constant. Then,

$$\delta/USY = \text{ const } = K. \qquad \ldots(5.5)$$

The diffusion coefficient is determined using the well-known equations given in [62].

Table 5.2 gives the results of determining the diffusion coefficient of two types of polymer concretes by different methods.

Table 5.2. Diffusion coefficients of FAM and PN polymer concretes

Polymer concrete	Method of determination	Diffusion coefficient $D \times 10^{-8}$ under the action of		
		10% H_2SO_4	10% NaOH	30% ammonium nitrate
FAM	Sorption	2.17	2.24	2.81
PN	Sorption	2.02	2.05	2.63
FAM	Amperometric	2.72	—	3.93
PN	Amperometric	2.26	—	2.12

The data shown in Table 5.2 reveal that the sorption method gives somewhat higher [sic] results compared to the amperometric method. This may be explained as due to the loss of a portion of the liquid absorbed by the specimen during its drying by filter paper. Moreover, the amperometric method is more accurate because of the continuity of the process. The diffusion coefficient of FAM polymer concrete is 10–15% more than that of PN polymer concrete and 25–30% more than that of FAED polymer concretes. On adding finely dispersed graphite to the polymer concrete, the diffusion coefficient generally decreases.

By knowing the diffusion coefficient, the time of liquid travel to a given depth of the material t_{dif} can be determined:

$$t_{dif} = \delta^2/\pi^2 D \; 3600. \qquad \ldots(5.6)$$

The time dependence of the advance of the diffusing liquid front can be described by the equation of [62]:

$$h_a = H(1 - e^{-\alpha t}), \qquad \ldots(5.7)$$

where h_a is the depth of penetration of the liquid in time t; H the specimen thickness on one-sided diffusion, cm; and α an empirical constant which contains the coefficient of penetrability (Table 5.3).

Table 5.3. Coefficient of penetrability $\alpha \times 10^{-9}$ for FA compositions based on furan resins relative to the type of the filler and medium, g/(cm-sec)

Aggressive medium	Andesite	Marshalite	Graphite
10% H_2SO_4 solution	0.84	—	—
10% NaOH solution	0.562	—	—
Saturated solutions of			
urea	—	0.177	0.143
ammonium nitrate	—	0.88	0.18
ammonium sulphate	—	0.51	0.78

5.3. Chemical Stability of Polymer Concretes in Aggressive Media

One of the most important properties of polymer concretes compared to ordinary concretes based on mineral binders is the high resistance of the former to the action of various aggressive media. Therefore, it is more rational to use products and structures based on polymer concretes under conditions of exposure to various aggressive media with no additional chemical [resistance] protection.

Innumerable investigations of the corrosion resistance of heavy and light polymer concretes of various compositions carried out in various organisations showed that, with an increasing concentration of acids within the range of their non-oxidising properties, the resistance coefficient of polymer concretes as a whole increases. Sulphuric acid in relation to polymer concretes is more aggressive than hydrochloric acid; maximum reduction of strength is noticed under the action of water on the polymer concrete (Table 5.4). This has been explained by the fact that, with an increasing concentration of the acid, the amount of water in the solution decreases and correspondingly the adsorptive strength reduction of polymer concretes.

The index of strength has been adopted as a decisive criterion for evaluating stability since it expresses a distinct association between the mechanical and physical-chemical properties of the material. The coefficient of stability K_{st} is determined as the ratio of the strength of polymer concretes after retention in aggressive media to the initial strength (σ_0):

$$K_{st} = \sigma_t/\sigma_0. \qquad \ldots (5.8)$$

Experimental investigations showed that the dependence between the reduction of strength and duration of test of the polymer concrete in aggressive media in rectilinear co-ordinates is curvilinear and may be described by the equation:

$$K_{st} = mt^b, \qquad \ldots (5.9)$$

where m and b are constants; and t the duration of retention in the aggressive medium, months.

In the zone of prolonged action of aggressive media, this dependence in logarithmic co-ordinates is persistently linear for all types of polymer concretes and aggressive media studied and is well approximated by the equation

$$\lg K_{st} = \alpha + b \lg t. \qquad \ldots (5.10)$$

Having calculated the coefficients $\alpha = \lg m$ and b from the experimental data, the durability of polymer concrete structures can be determined for a large time frame. However, the usual method of determining the corrosion resistance of polymer concretes is laborious since it requires the preparation of a large number of specimens. The non-destructive method of determining the stability coefficients from the change in the dynamic modulus of elasticity is more advanced and quite reliable:

$$K_d = E_{d \cdot t}/E_{d \cdot 0}, \qquad \ldots (5.11)$$

where $E_{d \cdot 0}$ and $E_{d \cdot t}$ are the dynamic modulus of elasticity initially and at moment t respectively.

The determination of the dynamic modulus of elasticity of polymer concrete is based on measuring the propagation of elastic waves which is associated with the modulus of elasticity as follows:

$$E_{d \cdot t} = V^2 \rho / K, \qquad \ldots (5.12)$$

where ρ symbol is the density of the material; and K a coefficient, $K = 1$ for a specimen of 40 mm $\times$ 40 mm $\times$ 160 mm.

Experimental investigations of the dynamic modulus of elasticity using the ultrasonic device DUK-20 showed that the dynamic coefficient of stability K_d obeys a dependence similar to that of K_{st} determined by the usual methods:

$$K_d = nt^k, \qquad \ldots (5.13)$$

$$\lg K_d = \lg n + K \lg t, \qquad \ldots (5.14)$$

where n and K are constants; and t the duration from the moment of commencing the tests.

Correlation analysis of the test results showed that the correlation coefficient r is close to one, i.e., the dependence $\lg K_d - \lg t$ forms a good straight line. Consequently, the dependence $\lg K_{st} - \lg K_d$ should also bear a rectilinear character. This has been confirmed in Fig. 5.3.

To determine the variation in strength of the material from the variation of dynamic modulus of elasticity, an equation should be derived relating K_d to the coefficient of stability K_{st},

$$\lg K_{st} = \lg K_d C_1 - C_2, \qquad \ldots (5.15)$$

where C_1 and C_2 are coefficients depending on the type of polymer concrete and acid concentration.

Table 5.4. Chemical stability of different types of polymer concretes

Aggressive medium	Concentration of medium, %	K_{st} (not below) at t = 20°C			
		FAM		FAED	
		dense	porous	dense	porous
Mineral acids:					
nitric	3	—	—	—	—
nitric	50	—	—	—	—
sulphuric	3	0.8	0.8	0.8	0.8
sulphuric	30	0.8	0.8	0.5	0.5
sulphuric	70	0.8	0.8	0.3	0.3
sulphuric	96	—	—	—	—
hydrochloric	5	0.8	0.8	0.8	0.8
hydrochloric	36	0.8	0.8	0.5	0.5
phosphoric	5	0.8	0.8	0.6	0.6
Organic acids:					
lactic	35	0.8	0.8	0.6	0.6
citric	10	0.8	0.8	0.6	0.6
acetic	5	0.7	0.7	0.6	0.6
Salts and bases:					
aqueous solution of ammonia	10	0.8	0.8	0.8	0.8
caustic soda	1	0.8	0.8	0.8	0.8
caustic soda	10	0.8	0.8	0.6	0.6
copper sulphate	5.3	0.8	0.8	0.8	0.8
Chloride solutions of salts of iron, calcium, magnesium, and sodium	Saturated	0.8	0.8	0.8	0.8
Solvents:					
acetone	100	0.7	0.7	0.7	0.7
benzene, toluene	100	0.8	0.8	0.8	0.8
ethyl alcohol	96	0.8	0.8	0.8	0.8
Petroleum products (diesel fuel, gasoline, kerosene, petroleum residue)	100	0.8	0.8	0.8	0.8

Note: "—" signifies that the use of the material in these media is not permissible.

Fig. 5.4 shows a graph of the variation of coefficients C_1 and C_2 for light polymer concretes based on polyester resin relative to the concentration of sulphuric acid plotted by mathematical processing of the data given in Fig. 5.3.

The following equations describe the variation of C_1 and C_2:

$$C_1 = 3.75 - 0.0583c, \qquad \ldots (5.16)$$
$$C_2 = 0.075 - 0.01165c, \qquad \ldots (5.17)$$

where c is the concentration of sulphuric acid, %.

after 12-month test period

for polymer concretes based on dense and porous aggregates

PN		KF–ZH		MMA	
dense	porous	dense	porous	dense	porous
0.5	0.5	—	—	0.8	0.8
—	—	—	—	—	—
0.8	0.8	0.8	0.8	0.8	0.8
0.8	0.8	—	—	0.8	0.8
0.5	0.5	—	—	0.5	0.5
—	—	—	—	—	—
0.8	0.8	0.8	0.8	0.8	0.8
0.8	0.8	—	—	0.8	0.8
0.8	0.8	0.8	0.8	0.8	0.8
0.8	0.8	0.5	0.5	0.8	0.8
0.8	0.8	0.5	0.5	0.8	0.8
—	—	—	—	0.8	0.8
0.6	0.6	0.6	0.6	0.8	0.8
0.8	0.8	0.8	0.8	0.8	0.8
0.6	0.6	—	—	0.8	0.8
0.8	0.8	0.8	0.8	0.8	0.8
0.8	0.8	0.6	0.6	0.8	0.8
0.8	0.8	0.8	0.8	—	—
0.8	0.8	0.8	0.8	0.7	0.7
0.8	0.8	0.8	0.8	0.8	0.8
0.8	0.8	0.8	0.8	0.8	0.8

Based on equations 5.16 and 5.17, graphs were plotted showing the changes of K_{st} in time (Fig. 5.5). The deviation between the data obtained using equation 5.15 and the experimental data does not exceed 3–5%, i.e., is commensurate with the experimental accuracy.

5.4. Biostability of Polymer Concretes

In addition to inorganic corrosive media, organic aggressive media, animal and plant organisms and their derivatives, i.e., organogenic aggressive media,

94

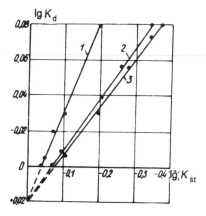

Fig. 5.3. Dependence of coefficient of stability K_{st} on dynamic coefficient of stability K_d

1—in 30% sulphuric acid solution; 2—in 10% sulphuric acid solution; and 3—in water.

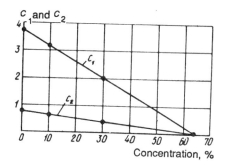

Fig. 5.4. Dependence of coefficients C_1 and C_2 on acid concentration.

exert an extremely destructive action on building materials and structures and installations. The present annual direct losses officially attributed to biogenic factors stand at about 2% of the production cost of industrial products [12].

Biological degradation of materials is known to be accompanied by the destructive action of organogenic aggressive media liberated by organisms and plants (root sap, excreta, excrements, etc.). The destructive action of organogenic aggressive media, in turn, is accompanied by active microorganisms and thus organogenic corrosion may be regarded as a variety of biological degradation. A particularly destructive action of biological organisms and organogenic aggressive media is manifest in food, light, chemical and medical industries, and in agricultural and hydraulic buildings and installations, etc. [24]. For example,

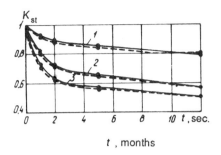

t , months

Fig. 5.5. Dependence of coefficient of stability on retention time in aggressive media. Continuous lines—experimental data; broken lines—calculated data:

1—in 30% sulphuric acid solution; 2—in 10% sulphuric acid solution; and 3—in water.

the asphalt-concrete floorings in the fat storage depots of meat combines are damaged in 2–3 months of use. The cement in floorings made of ceramic tiles is damaged in the fat and guts shops of meat combines as well as in sausage shops and the life of the tiles does not exceed 2.5–3 years.

Information is available that concrete partitioning structures and particularly the floorings in animal husbandry complexes are damaged within 3–4 years of use. Therefore, in the system of measures aimed at increasing the biostability of building materials and structures, this problem is quite important.

It has been established that organogenic aggressive media are highly corrosive of cement concrete. For example, cement concrete on daily contact with pig's fat heated to 70°C is damaged in 1–2 months. Cement concrete corrodes quite intensely on contact with vegetable oils, milk products, fruit and vegetable juices, and other edible products. The introduction of various additives including polymer emulsions and latexes improves the biostability of the cement concrete only insignificantly [24].

Apart from biochemical destructive phenomena, biomechanical, biophysical, and bioelectrical corrosion processes are also noticed in organogenic corrosion. The accompaniment of organogenic aggressive media by the growth of various microorganisms complicates the methods of improving biostability and calls for imparting appropriate bactericidal properties to the materials. Thus, the mechanisms of biological degradation and organogenic corrosion are more complex and multifaceted compared to the corrosion processes caused by inorganic aggressive media.

Studies have shown that polymer concretes based on furan, phenolformaldehyde, urea-formaldehyde, epoxide, and other resins possess utmost corrosion resistance against organogenic aggressive media. Polymer concretes based on these resins possess not only a far greater biostability compared to cement and asphalt concretes but also high bactericidal properties. Thus, a 2% aqueous solution of furfural totally suppresses fermentation while furfural, formaldehyde, and other constituents of polymer binder reduce or suppress bacterial growth.

Along with high biostability, some types of polymer concretes suffer from certain deficiencies: presence of free, exogenic, biologically active substances in many polymer binders which exert adverse chemical-biological effects on the biosphere and edible products; content of free water present in the binders which reduces the biostability of phenol and carbamide polymer concretes.

Thus, the following basic measures can be recommended for improving the operational characteristics including the biostability of polymer concretes: modifying mineral fillers with surface-active agents for improved wetting, workability, and a better ultimate density of polymer concretes and adding (when required) bactericides and fluorine-containing mineral fillers to the polymer concretes.

5.5. Corrosion Resistance of Reinforcing Metal in Polymer Concretes

It is known that steel reinforcement in dense reinforced concrete does not corrode for a very long time. Absence of steel corrosion in cement concretes is explained by steel passivation in the alkaline medium of such concretes. The main factor contributing to steel passivation in cement concretes is the constant contact of the steel with pore fluids having a pH value of 12, at which inhibition of the anode process is total. Corrosion of the reinforcing metal in cement concretes usually commences in carbonised concrete when the conditions are suitable for a fairly free anode metal dissolution process.

In polymer concretes the mechanism of protection and the possible processes of corrosion of reinforcement differ significantly from those observed in cement concretes.

Technical literature provides very little information on the corrosion resistance of reinforcing metal in polymer concretes and the information available is contradictory. Hence let us review, albeit briefly, the results of investigations carried out in this field.

Direct experiments on the corrosion resistance of reinforcing metal were carried out by the method developed at the Reinforced Concrete Research Institute on rods of steel St5 polished to Class 7 purity and degreased. Three such rods were fixed in each polymer concrete specimen with a cover (protective layer) of 7, 15, 20, 25, 30, and 35 mm thickness around them. The rods were 6 and 8 mm in diameter and 100 mm long. The protective layer (cover) of 7 mm thickness was used to facilitate quantitative assessment in a comparatively short duration.

The fully hardened polymer concrete specimens were tested in the most typical media: air-dry, atmospheric and aqueous solutions of acids (sulphuric and hydrochloric) and products typical of copper electrolysis. Tests were carried out at high temperatures in various liquid media.

During the tests the polymer concrete specimens were thoroughly inspected before breaking and recovering the reinforcing rods from them.

The condition of the steel reinforcement was evaluated from the following indexes: area of the spread of corrosion, visually as percentage of the area of the specimen; weight loss due to corrosion by weighing the rods in an analytical balance with an accuracy of up to 0.001 g; penetration depth of corrosion in μm determined in measuring microscope MIS-11 and when the pitting was deep by using an indicator with needle; and corrosion rate g/m^2 calculated as

$$V = \Delta G/St, \qquad \qquad \ldots (5.18)$$

where ΔG is the weight loss of the reinforcing metal specimen, g; S the surface area of the rod, m^2; and t the test duration, h.

All the indexes were determined simultaneously in not less than three parallel specimens of the reinforcing metal. Unlike steel, the corrosion stability of glass-plastic reinforcement was characterised by a coefficient of stability K_{st} which showed a change of strength on bending the glass-plastic reinforcing rods (diameter 6 and length 100 mm) after testing in polymer specimens R_t compared to the strength of the control specimens R_0:

$$K_{st} = R_t/R_0 \; . \qquad \qquad \ldots (5.19)$$

pH determinations in aqueous extracts of the soluble portion of various polymer concretes established that polymer concretes based on polyester and epoxide resins have a pH corresponding to that of neutral solutions. The pH of the soluble portion of polymer concrete based on furan resins is acidic. Further, with an increasing amount of benzenesulphonic acid added as hardener, pH decreases.

Direct tests in aggressive media revealed practically no corrosion of the reinforcement in all the test media except hot acids when used in polymer concretes based on polyester and epoxide resins at a thickness of over 20 mm. In hot acid solutions, which are capable of high penetration and are reactive to steel, corrosion was noticed. It roughly corresponded to the corrosion of the reinforcement in FAM polymer concretes tested in acid solutions at normal temperature (Table 5.5).

Table 5.5. pH Values of various polymer concretes

Polymer concrete based on	Amount of acidic hardener, %	pH
Binder PN-1	—	7–8
Binder ED	—	7–8
FAM	20	4.5–5
FAM	30	3.8–4

In FAM polymer concretes tested under air-dry conditions, corrosion of the reinforcing metal was not noticed. In liquid media, the steel reinforcement was affected independent of the type of the medium. However, the extent of corrosion damage varied very widely and depended on the following factors: composition of the polymer concrete, thickness of the protective layer, and the temperature of the aggressive medium.

Tests showed that, with an increase in thickness of the protective layer, corrosion of the steel reinforcement was significantly inhibited while the processes proceeded more intensely with an increase in temperature of the aggressive medium (Table 5.6) since the diffusion rate of the liquid media through the protective layer of the polymer concrete then increased.

Table 5.6. Corrosion of steel reinforcement in heavy FAM polymer concretes

Aggressive medium	Protective layer thickness, mm	Test duration, months	Corrosion area, %	Weight loss, g/m^2	Corrosion rate, $g/m^2 \times 10^4$	Maximum corrosion depth, μm
10% H_2SO_4 18–20°C	15	6	—	0.9	0	—
		18	30	13.3	10.4	36
	20	6	—	1	0	
		18	5	4.8	3.7	—
	30	6	—	0.9	0	—
		18	—	0.9	0	—
	35	6	—	1	0	—
		18	—	0.9	0	—
10% H_2SO_4 50°C	15	6	20	11.4	26.4	42
		18	70	27.3	18.5	88
	20	6	10	7.2	16.5	80
		18	30	22.6	17.8	90
	30	6	—	3.2	7.4	—
		18	25	14.2	13	86
	35	6	—	1.8	4.1	—
		18	20	10.8	10.5	70

In polymer concretes with acidic hardeners, the type of fillers and aggregates exerts a significant influence on the corrosion of the reinforcement in liquid aggressive media. Polymer concretes based on granite and andesite fillers and aggregates afford maximum protection to the reinforcing metal. The addition of porous aggregates to polymer concretes leads to some increased corrosion of the reinforcing metal.

Recognising the danger that corrosion of the reinforcing steel proceeds far more intensely in polymer concretes based on porous aggregates than in heavy polymer concretes, the test duration of the reinforcement in polymer concrete specimens based on porous aggregates was extended to 36 months. Under identical conditions of aggression, the rate of corrosion of the steel reinforcement in heat-treated specimens was, on average, 18–30% less at the protective layer thicknesses of 15 and 25 mm than in polymer concretes hardened under normal conditions (Table 5.7). The rate of corrosion of FAM polymer concretes based on agloporite rubble and silica sand is 1.5 times less than in FAM polymer concretes based on agloporite rubble and porous sand.

Table 5.7. Corrosion of steel reinforcement in light FAM polymer concrete based on agloporite

Aggressive medium	Hardening regime*	Protective layer thickness, mm	Penetrability, mm/year	Weight loss, g/m^2
10% H_2SO_4	N	15	0.0031	66
		25	0.0017	28
		35	0.0012	11.9
18–20°C	T	15	0.0027	53.3
		25	0.0012	19
30% H_2SO_4	N	15	0.0013	10.2
		25	0	0.46
		35	0	0.5
18–20°C	T	15	0.001	8.45
		25	0	0.5

*N—Normally hardened specimens; T—Heat-treated specimens.

When testing reinforced specimens of the first composition in 30% sulphuric acid solution for three years, faint traces of metal corrosion (about 0.5 g/m^2) were noticed on the reinforcing rods with a protective layer of 25 and 35 mm thickness. Such insignificant corrosion damage after three years of testing under extremely severe conditions convincingly demonstrated that FAM polymer concretes based on porous aggregates also ensure totally reliable protection of the reinforcing metal when the composition is correctly formulated.

Corrosion of steel reinforcement in polymer concretes was maximum when graphite was present in their composition although the stability of the polymer concretes themselves with graphite was extremely high. This is explained by the fact that graphite together with the reinforcing steel forms an electrochemical pair in which graphite functions as a cathode and steel as an active anode. Corrosion proceeds mainly as a result of the formation of local microelements on the surface of the reinforcement leading to intense pitting and ultimately to total destruction of the steel reinforcement.

Tests showed that under air-dry and atmospheric conditions, the polymer concrete media, including those using acidic hardeners, do not affect the strength of glass-plastic reinforcement. After 18 months of tests, there was practically no decrease in the value of K_{st}.

Tests of polymer concrete specimens with steel and glass-plastic reinforcement in liquid aggressive media established that polymer concrete specimens reinforced with steel undergo greater damage. This is explained by the discontinuities formed in the polymer concrete by the products of steel corrosion which are formed on the surface of the reinforcement. As their volume increases, they generate considerable splitting forces. After tests in acidic media, the strength reduction of glass-plastic reinforcement in polymer concrete specimens is less significant than under the action of alkaline media in cement concretes.

Direct corrosion tests established that the corrosion of the reinforcement commences some time (lag period) after the contact of the reinforced polymer concrete with the aggressive liquid. This duration is determined by using the coefficient of penetrability. After this, the rate of corrosion may be tentatively divided into three zones. In the first zone, an increase in corrosion rate is noticed as a result of the accumulation of the aggressive liquid by diffusion and an increase of its concentration on the metal surface. The second zone is characterised by a reduction of corrosion rate in time as a result of the growing electrochemical inhibition of the corrosion process because of the formation of a dense layer of corrosion products on the reinforcement and their screening of the metal surface. The third zone, mainly characteristic of polymer concretes with low stability and small thickness of the protective layer, represents the zone in which the corrosion rate again increases as a result of pitting damage of the reinforcement.

The second zone is not noticed in polymer concretes with graphite filler while the third zone is absent in polymer concretes based on granite aggregates and andesite filler. For such polymer concretes having a protective layer thickness of 30 mm and at a constant action of 10% sulphuric acid at 20°C, the corrosion rate in the second zone is roughly 0.003 to 0.004 mm/year. At this corrosion rate, the reinforcing metal can be used wholly reliably for at least 25 years.

For reinforced polymer concretes based on water-soluble phenol-formaldehyde and carbamide resins used under conditions of regular wetting, preliminary protection of the steel reinforcement with polymer coatings is necessary.

6

Durability of Polymer Concrete Structures

6.1. Modern Concepts of the Degradation Mechanism of Polymer Materials

A comparison of the experimental results of short- and long-term static strength, creep, plastic deformation, residual deformations, long-term strength of polymer concretes under repeated pulsating loads, etc. revealed the general patterns of all types of mechanical tests and loading regimes. For example, rapid loading to a definite level of stresses followed by unloading shows a direct dependence between stresses and deformations. In polymer concrete, residual deformations are not noticed, deformations dampen on prolonged application of comparatively small loads, and, at high loads, they initially dampen but later rise rapidly. It may be assumed that the commonness of these patterns is due to a common physical mechanism associated with the process of structural changes of the material under load.

The similarity of creep curves under prolonged static and repeated pulsating loads at different levels provides a basis for assuming that the degradation mechanism in both cases is of a common physical and mechanical nature. A knowledge of this makes it possible to explaining the time—deformation phenomena from a common viewpoint and enables a more confident prediction of material behaviour with allowance for the nature and history of loading.

Two types of polymers can be distinguished from the viewpoint of deformation mechanics [36, 43]:

Linear polymers (most thermoplastic polymer materials): In these materials, creep deformation is made up of (a) elastic, (b) high-elastic and (c) plastic deformations (Fig. 6.1, *a*). In this case, the bulk of total deformation is made up of high-elastic and plastic deformations while elastic deformation is perceptibly manifest only on rapid loading or at relatively low temperatures. Polymers of this type do not exhibit equilibrium deformation since, under the action of constant external loads, high-elastic deformations dampen exponentially while plastic deformations increase irreversibly proportional to the loading time. On

102

unloading, the high-elastic deformations recover in time while the value of residual deformations is determined by the plastic deformation.

Spatially cross-linked polymers (most thermosetting resins): There is a view that the macromolecules of these polymers have a common spatial cross-linkage and that the polymer product essentially represents one single gigantic molecule. If this is so, creep deformations in these polymers can be seen only as a result of elastic and high-elastic deformations. For polymers of this type only reversible deformations and absence of residual deformations after unloading should be a characteristic feature (Fig. 6.1, *b*).

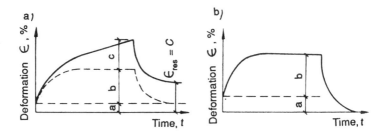

Fig. 6.1. Typical deformation curves of (a) thermoplastic and (b) thermosetting polymers.

Based on these premises, some scientists point out that the equilibrium and plastic deformations are mutually exclusive; the material has either equilibrium deformation with no plastic deformation or residual deformations do not arise (spatially cross-linked polymers) or viscous flow occurs and residual deformations are caused by it in the material but no equilibrium deformation (linear polymers). The manifestation of residual deformations in cross-linked polymers can be explained only by partial damage of the cross-linkages.

However, even theoretically it is difficult to assume the possibility of a total spatial cross-linkage of all the macromolecules in a polymer block. In real systems, the degree of polymerisation of oligomers in the best case is 97–98%, but in practice it does not exceed 95–96%. In large-sized products characterised by significant temperature drops along the section during hardening, the overall degree of polymerisation is even low. It should further be added that hardening of many thermosetting resins occurs in the presence of catalysts which do not form a chemical compound with the polymer molecules but remain in the body of the material.

It is quite natural to assume, therefore, that in such systems the spatially cross-linked molecules may alternate with unreacted molecules of the oligomer and catalyst inclusions. In that case, the cross-linked polymers may undergo limited plastic deformations without damage of the cross-linkages while their values are determined by the volume of the viscous phase in the polymer block.

Actual degradation of most of these materials is caused by the gradual onset and development of microfissures. These aspects have been studied by P.A. Rebinder, I.N. Akhverdov, S.N. Zhurkov, M.I. Bessonov, N.V. Mikhailov, A.I. Slutsker, V.S. Kuksenko, V.E. Gul', and others. Considerable experimental data have been accumulated on this subject and on this basis various hypotheses have been posited to explain the degradation kinetics of materials. However, there is presently no unanimity regarding the physical principle of degradation [2, 13, 57].

For example, optical microscopy [36] established that microfissures are formed throughout the polymer body sometime after its loading, independent of the load applied. On holding the specimen under load, new fissures form while the old ones grow.

The formation of microfissures is an irreversible process. Prolonged resting of the material after releasing the load does not 'heal' the fissures and they begin to grow again after the next loading.

Interesting investigations on the formation and growth kinetics of submicroscopic fissures were carried out by S.N. Zhurkov, V.S. Kuksenko, and A.I. Slutsker. Using the method of X-ray scattering at low angles, extremely fine fissures were detected in polymers under uniaxial tension. These submicroscopic fissures were in the form of disks ranging in diameter from a few tens to hundreds of ångströms with their planes oriented perpendicular to the loading axis. It was experimentally shown that submicroscopic fissures arise extremely unevenly under stress: initially, the rate of their formation is very high but gradually dampens in the course of time; as the load increases, the rate of formation of such fissures increases exponentially.

It was also established that the overall concentration of fissures before damage, independent of the type of polymer and the load, is practically the same—from 10^{14} to 10^{15} cm^{-3}. At this concentration and size of detected fissures, their proximation is such that $a/L \leq 3$, where a is the distance between the closest ends of fissures and L the fissure length (diameter).

Theoretical calculations have shown [5] that under the action of tensile stresses at $a/L < 3$, fine fissures merge into much larger ones by the breakdown of barriers between them. This process of merging of fissures takes place at random at some place where two fissures located close to each other begin to grow rapidly, resulting in the formation of a major fissure and fracture of the specimen.

Unfortunately, the authors [5] carried out all the tests mainly on polymers belonging to the first group at comparatively high stresses (60 to 200 MPa). The nature of the origin and growth of fissures in spatially cross-linked polymers at low stresses was not adequately studied.

While studying degradation kinetics of hardened unfilled thermosetting resins under the action of shrinkage stresses, we have also detected lamellar micro- and macrofissures in the form of regular disks which, as they grew, could be detected by the naked eye. In all probability, at an appropriate stress level, the degradation mechanism of such polymers would be the same as that described in many works.

At the same time, it must be pointed out that almost no investigations have been carried out on the degradation mechanism in highly filled polymer materials based on thermosetting resins.

6.2. Combined Action of Synthetic Binders and Fillers

High values of physical and mechanical properties of filled polymer compositions can be obtained only when the strength of the adhesive bonds of the synthetic binder with the surface of fillers is adequate. The strength of the adhesive bonds of many thermosetting resins with the surface of mineral fillers (silicates, carbonates, etc.) may go up to 10 to MPa [5, 47]. On this basis, some authors point out that when the stresses in the binder exceed the strength of the adhesive bonds, the polymer shell-matrix peels off from the filler surface and then the ultimate tensile stresses depend only on the stresses arising in the remnant polymer portion. However, experimental investigations of the growth kinetics of deformations and the nature of degradation under different loading regimes do not confirm these assumptions.

The $\sigma - \epsilon$ diagrams plotted by a loop oscillograph on rapid loading of the specimen at above 60 MPa/min confirm a linear dependence between σ and ϵ up to a certain level of stresses ($\sigma < R_k$). At $\sigma < R_k$, after the first cycle of rapid loading and unloading, insignificant residual deformations were noticed; later, on repeating such cycles seven times, residual deformations were not noticed (Fig. 6.2). This provides justification for assuming that under such loading regimes, the adhesion bonds between the binder and the filler are not broken down and the filler, taking upon itself the corresponding load, will deform together with the polymer shell-matrix in proportion to its modulus of elasticity (for example, for polyester resin PN-1, the modulus of elasticity is roughly 2.5×10^3 and for silica sand 5×10^4 MPa).

Investigations carried out using optical and electron microscopes on specimens filled with such stable fillers as quartz, granite, andesite, etc., did not reveal the breakdown of adhesion bonds of the binder with the filler surface. In all cases, during tests under uniaxial tension, diagonal tension, bending and compression, breakdown occurred with a rupture of the fillers and along the critical locations in the body of the polymer. Further, the undamaged particles of the filler emerging onto the fracture surface are invariably covered with a polymer shell.

Thus, direct experiments did not confirm the prevailing view and showed that under any stresses arising in the filled polymer compositions including polymer concretes, such systems function as a single unit right until the point of breakdown, without damage of the adhesive bonds between the binder and

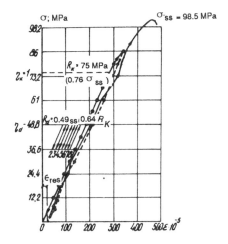

Fig. 6.2. Deformations on short-term loading and unloading (oscillogram data). ss—strength of shell

the filler and aggregate grains. Consequently, the strength of adhesive bonds in many types of polymer concretes is invariably greater than the strength of the most frequently used fillers and aggregates and the cohesive strength of the polymer. These premises agree well with the data of [11] and many other researchers.

The conditions for the breakdown of bonds between the polymer molecules and the subsequent formation of submicroscopic and microscopic fissures in filled polymer compositions based on thermosetting resins differ significantly from the conditions characteristic of polymers of the first group or unfilled thermosetting plastics. In this case, the rupture of chemical bonds and the formation of submicroscopic fissures are noticed primarily at the critical locations in the polymer body at some distance from the filler surface. The merger of minute fissures into much larger ones due to the breakdown of barriers between them will arise under more complex conditions since, sooner or later, the apexes of these fissures fall in the zone of more orderly structures close to the filler surface. Further growth of such fissures can arise either around the filler surface or through the rupture of the polymer 'shell' and the filler itself (Fig. 6.3). This has been confirmed by direct observations of the nature of the fracturing of filled compositions. In the first as well as in the second case, the work expended on rupture in the filler zone increases, which serves as an additional proof of the effect of reinforcement of polymers with mineral fillers.

Thus, under the influence of externally applied forces, the deformation of filled polymer compositions based on thermosetting resins can be represented as follows. At the first moment of applying the load, elastic deformations arise in the material followed by limited plastic deformations, which are proportional to the

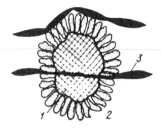

Fig. 6.3. Sketch showing development of fissures in the zone of a filler grain:

1—filler grain; 2—supramolecular formations of adhesive layer; and 3—micro- and macrofissures.

quantum of the viscous phase, and high-elastic deformations (Fig. 6.4). As long as plastic deformations do not cease, it is difficult to assume the development of microfissures in the material. If such microfissures do arise, they will be of sublocal character and will not be responsible for the behaviour of the material under load. Only after the cessation of plastic deformations can the phenomenon of microfissures be seen in the material. On attaining the critical value, these microfissures lead to the degradation of the material.

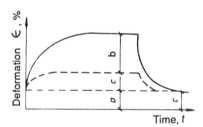

Fig. 6.4. Characteristic curve of deformation of polymer concretes based on thermosetting resins:

a—elastic deformation; b—high-elastic deformation; c—limited plastic deformation.

Since plastic deformations are irreversible, two consequences are possible: after unloading, polymer concrete specimens may have residual deformations not associated with the processes of microfissures; on subsequent compressive or bending loading of such specimens, the effect of compaction and some stiffening should be noticed and, under identical repeated loads, deformations of these specimens should be lower in value than those of plastic deformation.

Experimental investigations quite convincingly confirm the proposed deformation mechanism and the degradation process in polymer concretes.

Deformation studies on polymer concrete prisms 100 mm × 100 mm ×

400 mm in stage-wise compressive tests carried out by F.A. Luchinina and G.K. Solov'ev showed that, even under comparatively small loads, the effect of compaction is quite distinctly manifest. The dependence of the change of volume

$$\Delta Q = [-\Delta \epsilon_{i.long}(2\Delta \epsilon_{i.trans})]10^{-5} \qquad \ldots(6.1)$$

on the load applied in the process of compaction is shown in Fig. 6.5, a (here, $\Delta \epsilon_{i.trans}$ and $\Delta \epsilon_{i.long}$ represent the increment of total transverse and longitudinal deformations).

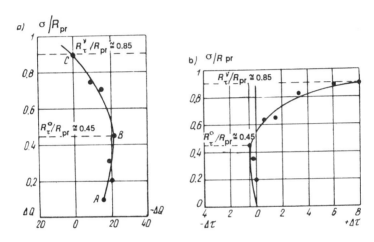

Fig. 6.5. Determining the boundaries of microfissures in FAM polymer concretes based on agloporite aggregates:

a—change of volume; b—change in propagation period of ultrasound τ, % of initial value.

In this case, the segment of curve AB strikingly reflects the reduction of polymer concrete volume, i.e., compaction of the material, and later the appearance and growth of microfissures leads to an increase of the volume (segment BC).

Further, the practically complete agreement between R_τ^o and R_τ^v determined from the propagation velocity of ultrasound and from the change in the volume ΔQ of polymer concrete specimens during the tests is a characteristic feature (Fig. 6.5, b).

In studies on the long-term strength and deformability of two-layer beams whose upper portion is made up of cement concrete and the lower portion of polymer concretes based on epoxide and polyester resins, V.V. Fridman noticed that, on repeated loading of such beams, total deformations are reduced [42]. Thus, experimental data recorded at the Reinforced Concrete Research Institute and the investigations of other scientists [32] convincingly reflect the occurrence of two phenomena in polymer concretes under load. Initially, the material undergoes compaction mainly under plastic deformation and later, at a certain level of stresses, microfissures appear and grow. These microfissures lead to an increase in the volume of the material and its subsequent destruction.

6.3. Temperature—Time Dependence of Long-term Strength of Polymer Concretes

The use of polymer concretes as load-bearing structures of industrial buildings and installations exposed to various aggressive media is not only rational and economically justified, but in some cases is the only possible solution for meeting the requirements. At the same time, extensive application of polymer concretes in construction practices is impeded by several factors, the most important of which is the lack of a reliable and practically convenient method of evaluating the long-term strength and deformation characteristics of polymer concretes and methods for designing load-bearing structures based on them. The specific properties of polymer materials render difficult the development of a common theory and methods for calculating the long-term strength and deformability of structural members based on them [10, 21, 43, 54].

The actual strength of many engineering materials is known to be a few orders less than the theoretical strength calculated on the basis of intermolecular forces. An explanation to this deviation was first offered by Griffith [34], who suggested the presence of rudimentary fissures in the solid body and demonstrated that stresses at their apexes coincide with the theoretical strength of the material while the average stresses applied to the material at the moment of failure represent a measure of its inherent strength.

Based on these concepts, a statistical theory of brittle fracture was proposed by Griffith [34]. However, this theory did not explain the dependence of strength on the duration of the effect of load and the critical nature of the breakdown. It was later shown that polymer materials break down as the result of a process continuously proceeding in time while the duration of total breakdown depends to a large extent on the duration of the effect of load and the temperature of the medium.

To determine the long-term strength of polymer materials with brittle fracture, Zhurkov [85, 86] proposed the following equation:

$$t = t_0 e^{\frac{U_0 - \gamma\sigma}{KT}} \qquad \ldots (6.2)$$

from which, at constant temperature,

$$\sigma = \sigma_0 - \text{const } t/t_0, \qquad \ldots (6.3)$$

where t is the duration in seconds before failure of the materials under stress σ; U_0 and t_0 constants characteristic of the material (Table 6.1); γ a structural coefficient; T absolute temperature; K the universal gas constant; and σ_0 the absolute ultimate strength of the polymer material, $\sigma_0 = U_0/\gamma$.

Equation (6.3) shows that strength depends significantly on the duration of the effect of load. The longer this duration, the smaller the ultimate strength. It follows from equation (6.2) that as the temperature increases, the strength of the polymers decreases. For a given duration of the effect of the applied force, the effect of temperature is linear in character to a certain extent:

$$\sigma = \sigma_0 - \text{const } T_0. \qquad \ldots (6.4)$$

Although equation (6.2) helps to eliminate contradictions characteristic of the statistical theory of strength, it is not devoid of some deficiencies. At $\sigma \to 0$, the long-term strength of the specimen is limited and, at $\sigma = \sigma_0 = U_0/T$, the long-term strength is independent of temperature, which contradicts the physical concept of equation (6.2) [37]. A theoretical relationship is given in [37, 62] which eliminates the physical paradoxes in equation (6.2).

Innumerable experimental investigations have shown that the value of t_0 lies in the range 10^{-12} to 10^{-13} (Table 6.1). This value is extremely proximate to the period of atomic oscillations in a solid body. Therefore, when predicting the long-term strength, constant t_0 cannot generally be determined. From this emerges an extremely important conclusion that the long-term strength, other conditions remaining comparable, is characterised by only two coefficients, U_0 and γ, which may be regarded as constants of a given material. Hence, for predicting the durability of polymer concretes, these coefficients should be determined experimentally (Table 6.1).

Table 6.1. Constants of polymer materials and polymer concretes based on PN-1

Material	t_0, sec	U_0, kJ/mole	γ, kJ $\times$ (mole $\times$ N)
Caprone fibre	10^{-12}	189	18
Polypropylene	10^{-13}	235	26.9
Polystyrene	10^{-13}	235	—
Polyvinyl chloride	10^{-13}	202	24.8
Polymethyl methacrylate	10^{-13}	130	—
Polymer concrete based on PN-1	10^{-13}	149	440–490

The above equations characterising the long-term strength are true only when all the physical and chemical transformations in the test materials have ceased and they possess stable properties.

In the course of hardening large-sized polymer concrete structures and products, significant temperature and ultimate internal shrinkage stresses arise in the material and damage its impermeability in some cases. The position becomes aggravated when such structures are used under conditions of the action of var-

110

ious aggressive media which, in turn, damage the structural stability state of the material. Therefore, a more detailed study of the physical principle of the mathematical equation of temperature—time dependence of the strength of such materials is essential.

Under stable temperature and humidity conditions, each type of material and the stress state of all the members figuring in equation (6.2), except the structural coefficient γ, have positive real values. The structural coefficient $\gamma = \beta\omega$ (ω is the fluctuating volume of the elementary act of breakdown and β the coefficient of stress concentration) characterises the structural state of the material and the uniform distribution of stresses in the external field of forces. Coefficient β, in turn, depends not only on the temperature, but also on the stress applied since these parameters determine the rate of relaxation processes. Thus, the value of coefficient β changes even under stable conditions.

For example, the initial hardening of actual field structures made of polymer concretes occurs in many cases at normal temperature of the environment, i.e., under isothermal conditions, with the liberation of a significant amount of heat into the ambient atmosphere. In this case, at the initial moment there is a uniform temperature field θ_0 (Fig. 6.6, a) throughout the section of the product. With the commencement of the hardening process, as a result of exothermal self-heating, heat transfer with the environment, and the relatively low thermal conductivity of polymer concretes, the temperature field in the section becomes uneven with a maximum at the centre of the section (Fig. 6.6, b). As the polymerisation (polycondensation) processes advance, this effect is intensified and the temperature difference Δt between the outer surface of the article and its centre increases rapidly (Fig. 6.6, c).

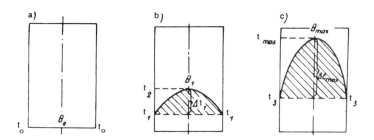

Fig. 6.6. Distribution of self-heating temperature along section of the product.

With an increase in temperature, the product should expand as a result of thermal deformations. The formation of chemical bonds during polymerisation is accompanied by the development of shrinkage deformations. Quite naturally, it was assumed that the diagrams of internal stresses at the moment of achieving maximum temperatures will appear as shown in Fig. 6.7 and the thermal stresses should be compensated by shrinkage stresses. However, experience points out that damage occurs in several cases under the influence of temperature stresses.

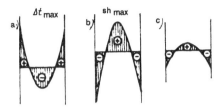

Fig. 6.7. Diagrams of internal stresses along section of the product: (a) under action of temperature deformations; (b) under action of shrinkage deformations; (c) overall diagram of internal stresses.

It has been established [62] that shrinkage deformations retard, and lag behind in time from, temperature deformations. Extremely important consequences emerge from this: (1) shrinkage deformations advance somewhat later than thermal deformations; (2) at high temperatures of exothermal self-heating, thermal deformations and the internal stresses accompanying them can damage the impermeability of the product and (3) the time differences between thermal and internal shrinkage stresses help to determine them individually.

Thus, high temperatures and thermal stresses accompanying them should not be permitted in the actual structures during their moulding and hardening. The maximum possible thermal stresses should be predetermined for each product or structure before commencing its production.

The non-linear distribution of temperature along the section of the product, the variable modulus of elasticity and the coefficient of thermal deformations, and the presence of relaxation processes complicate the method of calculation so much that it is practically impossible to use it for calculating thermal stresses in polymer concrete structures.

The modulus of elasticity of polymer concretes is known to depend significantly on temperature and can be described by the exponential relationship:

$$E_t = E(1 - e^{\alpha t}), \qquad \ldots (6.5)$$

where E is the modulus of elasticity of the polymer concrete at normal temperature; α a coefficient equal to 0.055 for polyester and furan resins; and t the temperature of polymer concrete.

Investigations have shown that in the range zero to $100°C$, the short-term strength and modulus of elasticity of polymer concrete prisms decrease in proportion to the increase in temperature. Consequently, in the course of hardening of polymer concretes, the maximum value of modulus of elasticity will be characterised by the self-heating temperature and cannot exceed the corresponding modulus of elasticity for the hardened specimens.

As the temperature varies from 20 to 100°C, the coefficient of thermal deformations of polymer concrete also changes linearly from 19×10^{-4} to 13×10^{-4}.

Based on the statistical processing of test results, a comparatively simple equation for determining the maximum thermal stresses has been proposed as under with an accuracy adequate for practical purposes:

$$\sigma_t = \frac{(E_0 - Kt_{max})10^3(\alpha t_{max} - \alpha t)}{1 - \mu} \psi, \qquad \ldots (6.6)$$

where σ_t are the maximum thermal stresses; E_0 the modulus of elasticity at 0°C; K the proportionality coefficient equal to $0.9 \ Pa \times °C$; t_1 and t_{max} the self-heating temperature on the surface and at the centre of the section respectively; α the coefficient of thermal deformations; μ Poisson's ratio, $\mu = 0.22 \ldots 0.275$; and ψ the coefficient of relaxation of thermal stresses, $\psi = 0.6 \ldots 0.7$.

Thus, changes in the structural state of polymer concretes can be studied only if coefficient β reflects the algebraic sum of all the stresses arising in the material:

$$\beta = \beta_1 + \sigma_t + \sigma_{sh} + \sigma_m, \qquad \ldots (6.7)$$

where β is the overall coefficient of stress concentration; β_1 the stresses due to external loads; σ_t, σ_{sk} and σ_m the thermal, shrinkage, and moisture stresses, respectively.

Consequently, structural coefficient γ should include the overall influence of coefficient β.

The low homogeneity and the presence of considerable inherent stresses can be explained in all probability by the fact that coefficient γ for polymer concretes is more than 10 times greater than the corresponding values for unfilled thermoplastic materials (Table 6.1).

Equation (6.2) can be written as

$$\sigma = U_0/\gamma - (KT/\gamma)[\ln (t/t_0)]. \qquad \ldots (6.8)$$

At $T = T_i$ = const, it follows from equation (6.8) that the stresses prevailing in the material are linearly associated with the logarithm of durability of the material.

Thus, to predict the durability of material under conditions of the combined action of loads and temperatures, U_0 and γ would have to be determined.

From the viewpoint of kinetic concepts of strength, the breakdown process under the combined action of loads and aggressive medium should be regarded as a process developing in the material with the passage of time. Therefore, the rate of accumulation of microfissures due to load V_n and the action of aggressive medium V_a can be regarded as the characteristics of damage. The resultant rate of damage V_s at any moment of time t is determined as a sum in the first approximation:

$$V_{s(t)} = V_{n(t)} + V_{a(t)}, \qquad \ldots (6.9)$$

where $V_{n(t)}$ is the rate of damage of the material under the action of load; and $V_{a(t)}$ the rate of damage of the material under the action of the aggressive medium.

The assumption that the resultant rate of damage is determined by the sum of the rates of each effect individually is extremely simple and has been confirmed by several experimental data [36].

For appropriate resultant estimation of durability (t_s), the following equations are relevant:

$$1/t_s = 1/t_n + 1/t_a \text{ or } t_s = t_a t_n / (t_a + t_n). \qquad \ldots (6.10)$$

The durability t_n of polymer concretes under the action of loads and environmental temperature, under tension and bending, was experimentally determined in specially designed equipment. By varying the load and also the temperature with the help of special devices, the change of durability can be traced over a wide range from a few seconds to months [62].

The experimental data obtained are shown in Fig. 6.8 in co-ordinates $\lg t - \sigma$. The mathematical processing of these data point to a linear relation under tension as well as bending. The correlation equations for the polymer concretes under consideration are shown in Table 6.2. The extrapolation of relationship $\lg t - \sigma$ for polymer concrete KF–Zh shows that all the three straight lines intersect at one point with the ordinate $\lg t = 13$ or $t_0 = 10^{-13}$ sec; this does not contradict the available data (Fig. 6.9).

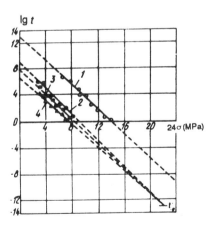

Fig. 6.8. Change of durability relative to stresses and temperature:

1—light polymer concrete PN-1 at 20°C; and 2, 3, and 4—light polymer concrete KF–Zh at 20, 60, and 80°C respectively.

114

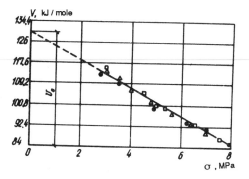

Fig. 6.9. Dependence of activation energy on stresses for polymer concretes KF–Zh:

• —results were obtained at 20°C; □ and △—results were obtained at 60 and 80°C respectively.

Table 6.2. Main constants of polymer concretes investigated

Index	Polymer concrete based on	
	KF–Zh resin (under tension)	PN-1 resin (under bending)
Correlation equation for:		
t = 20°C	$\lg t = 8.9 \ldots 0.98\sigma$	$\lg t = 12.96 \ldots 0.8$
t = 60°C	$\lg t = 7.8 \ldots 0.93\sigma$	—
t = 80°C	$\lg t = 6.56 \ldots 0.87\sigma$	—
Energy barrier U_0, kJ/mole	130	147
t_0, sec	10^{-13}	10^{-13}
Coefficient γ, kJ/mole × MPa	6.1	4.2

Knowing that $t_0 = 10^{-13}$ sec, U_0 and γ appearing in equation (6.2) can be determined.

It follows from equation (6.2) that

$$U_0 = \gamma\sigma = 2.303 KT(\lg t - \lg t_0). \qquad \ldots (6.11)$$

The values appearing on the right side of equation (6.11) are known and hence σ can be calculated for different values of $U_0 = \gamma\sigma$. The results of calculations using equation (6.11) for compositions based on KF–Zh are shown in Fig. 6.9. The values found at different temperatures fall quite well on a straight line corresponding to the equation $U = U_0 = \gamma\sigma$, from which U_0 and γ can be calculated (Table 6.2). Having obtained the values of t_0, U_0, and γ, durability t_n can be determined using equation (6.10).

Under the action of aggressive media, most polymer materials lose their strength over the course of time. The strength variation of polymer concrete (tensile under bending) in relation to the duration of exposure to aggressive media t_a can be represented as:

$$\sigma_a/\sigma_0 = mt_a^n, \qquad \ldots (6.12)$$

where σ_0 is the nominal strength at stress rate 0.1 to 0.2 MPa/sec; t_a the duration of exposure to the aggressive medium, sec; n and m constants depending on the type of the material and the aggressive medium; and σ_a the strength of the material after exposure to the aggressive medium for time t, MPa. It may be noted that equation (6.12) is applicable at $t_a \geq 30$ days.

The durability of polymer concrete under the action of aggressive media and in the absence of external loads may be expressed using equation (6.12) as

$$t_a = \sqrt[n]{\sigma_a/m\sigma_0} \ . \qquad \ldots (6.13)$$

This equation helps to determine the durability of the material for a given permissible reduction of its strength.

Constants m and n were determined for polymer concretes based on KF–Zh and PN-1 under the action of water and 10% and 30% sulphuric acid solutions (Table 6.3).

Table 6.3. Variation of coefficients n and m relative to the type of aggressive medium

Polymer concrete	Aggressive medium	Coefficients	
		n	m
PN-1 (bending)	10% H_2SO_4	−0.116	4.6
	30% H_2SO_4	−0.04	1.73
	H_2O	−0.183	11.73
KF–Zh (tensile)	10% H_2SO_4	−0.106	4.18
	30% H_2SO_4	−0.143	6.89
	H_2O	−0.089	2.94

Having determined durability in relation to external load t_n and the action of aggressive medium t_a, graphs were plotted showing overall durability (Fig. 6.10).

An analysis of the relationships obtained revealed breaks at some levels of stresses σ_{br} in the graphs of overall durability. In the zone of $\sigma > \sigma_{br}$, mechanical loads exert a decisive influence on durability since $t_n \ll t_a$, while in the zone $\sigma < \sigma_{br}$ the aggressive medium exerts a decisive influence on the overall durability since $t_n \gg t_a$.

The proposed method was used to calculate the durability of polymer concrete columns of acid bath platforms in copper electrolysis shops subjected to the combined action of loads, high temperatures, and spillages of sulphuric acid solutions. Long-time operational experience of load-bearing polymer concrete structures confirms the correctness of using this method for predicting similar structures made of various types of polymer concretes.

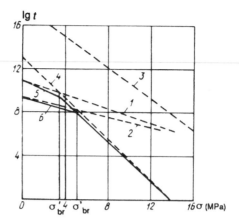

Fig. 6.10. Variation of durability of polymer concrete PN-1:

1—in 10% H_2SO_4; 2—water; 3—in 30% H_2SO_4; [4—not given in Russian text]; 5—under the combined action of mechanical load and 10% H_2SO_4; 6—under the combined action of mechanical load and water.

Calculations made by the proposed method show that the predicted durability of polymer concrete structures relative to the type of polymer concrete and the operational conditions can vary from 20 to 40 years.

The investigations discussed above and the significant fundamental studies of A.I. Chebanenko have helped to establish the distinctly inherent relaxation and hysteresis properties of polymer concretes caused by active man-made elasticity of the polymer matrix. These investigations formed the basis of the theory of designing reinforced polymer structures (see Chapter 8).

7

Effect of Temperature and Fire on Polymer Concrete Structures

Along with several positive properties, polymer materials and the plastics based on them suffer from many vital deficiencies, amongst which are the low thermal stability and the combustibility of many of the polymer compositions produced. The extent of filling of plastics with various inorganic materials is generally 30–60% by weight. Naturally, the greater the extent of filling and, correspondingly, the smaller the amount of polymer in the mix, the greater the increase in thermal stability and the lesser the possibility of inflammation of such a material. But, even at this level of filling most of the plastics burn quite well.

The inadequate attention paid to the combustibility and heat stability of plastics and, as a consequence, the failure of fire-fighting measures in the design stage as well as at the time of fabrication and use, has been the main reason for major fires in industrial, administrative, cultural, and residential buildings and installations in many countries, resulting not only in huge material losses but even loss of human life in several instances.

As pointed out earlier, polymer concretes contain 5 to 10% polymer binder and 90–95% incombustible mineral aggregates and fillers. Thus, the very composition of these concretes minimise their inflammability and combustibility.

The synthetic monomers and oligomers used most often as binders for producing polymer concretes have different ignition temperatures and specific heat of combustion which can differ by 1.5 to 2 times depending on the type of polymer used (Table 7.1).

The use of furan, carbamide, and phenol-formaldehyde resins as binders, since they possess a comparatively high ignition temperature, suggests that structures of polymer concretes based on them will possess a fairly high heat stability.

Considerable data has been reported in recent years on the problem of combustibility and heat stability of polymer materials and the plastics based on them. Investigations in the field of destruction and combustion of polymer materials, reduction of their combustibility and development of methods for increasing the

Table 7.1. Ignition temperature and specific heat of combustion of unfilled hardened polymers

Polymer	Ignition temperature, °C	Specific heat of combustion, kJ/kg
Polyester maleinate	250	29,300
Polymethyl methacrylate	280–300	26,400
Epoxide	400	—
Polyurethane	400–440	24,360
Phenol-formaldehyde	500	21,000
Urea-formaldehyde	450	17,640
Furfural-acetone	400–450	24,360

heat stability of building structures [3, 33, 66, 82] have shown that a characteristic feature of the combustion of polymer building materials is the multistage process of their conversion into the end products of combustion. Based on an analysis of these investigations, the combustion of polymer building materials can be regarded as a continuous process consisting of several stages: accumulation of thermal energy from the sources of ignition, decomposition of materials, ignition and combustion of volatile products of pyrolysis.

The principles of pyrolysis and combustion of polymers identified to date enable determination of possible methods of reducing their combustibility and increasing heat stability by inhibiting the reactions in the stage of pyrolysis, reducing heat transfer in the bulk of the mix, and inhibiting the combustion process itself. This can be achieved by introducing fire-proofing compounds and incombustible fillers and by chemical modification of polymers.

Investigations have shown that the most effective fire-proofing compounds for reducing the combustibility of polymer building materials without significantly reducing the operational indexes are the phosphorus-containing reagents. The mechanism of the action of these compounds involves an increase in the thermal-oxidation stability of polymers which is associated with reducing the amount of hot volatile products of destruction and increasing the yield of coke residue, which prevents heat and mass transfer during combustion [33]. It has been shown [82] that the phosphorus-containing reactive fire-proofing compound phosphocrylate, on addition to polyester resin PN-1 not only significantly reduces combustibility, but also promotes thermal stability of the hardened polymer.

Fire-proofing compounds and the objective modification of polymer building materials can be successfully adopted for monomers as well as oligomers used as binders in polymer concretes. These, together with a high filler content, ensure their maximum efficiency.

The results of investigations on the combustibility of polymer building materials can be used in formulating polymer concrete mixes at the stage of selecting the type of binder. At the same time, it should be pointed out that information on the combustibility and heat stability of polymer concretes is extremely scant in the Russian as well as foreign literature. Therefore, the Reinforced Concrete Research Institute, V.A. Kucherenko Order of the Red Banner of the Labour Central Research Institute of Building Structures, All-Union Research Institute of the Refractory Industry, Order of Lenin and the Red Banner of the Labour Moscow Institute of Railway Transport Engineers, and many other organisations are developing special methods and are also carrying out investigations on the effect of temperature and heat on polymer concretes and the structures based on them.

7.1. Effect of Temperature on Strength and Initial Modulus of Elasticity

The effect of temperature on strength and initial modulus of elasticity of heavy FAM polymer concretes was determined at the Reinforced Concrete Research Institute by carrying out studies on prism samples 70 mm × 70 mm × 280 mm and 100 mm × 100 mm × 400 mm at 20, 40, 60 and 100°C. Three specimens of each size were tested at each temperature.

A gentle rise of temperature was effected at 20°C/h in special muffle furnaces using a voltage regulator. For uniform heating all along the section, the prisms were held at the given temperature for 4h. The temperature distribution along the height of the prism was controlled by three thermocouples connected to a potentiometer.

Deformations were measured on a base of 100 mm using gauge-type indicators (having a least count of 0.01 mm) marked on the four sides of the prism, using special frames and extension devices.

Tests showed that in the temperature range up to 100°C, the ultimate strength and modulus of elasticity decreased proportional to an increase in temperature. Further increase in temperature caused a more intense reduction of strength and modulus of elasticity associated with the commencement of the thermal destruction of the polymer binder. Up to 100°C, the reduction of strength and rigidity of FAM polymer concretes was a reversible process, i.e., on reducing the temperature to 20°C, the strength and modulus of elasticity reverted to the original level (Fig. 7.1).

Increasing the retention period of specimens at 60°C to 100h caused almost no change in the established strength and modulus of elasticity of FAM polymer concretes.

These investigations enabled determination of coefficients of reduction of ultimate strength and modulus of elasticity of polymer concretes depending on temperature under short-term effect of loads (Table 7.2).

The strength of polymer concrete specimens heated to 150, 200, 300 and 400°C followed by cooling to 20°C was 0.8, 0.6, 0.36 and 0.2 R_{st} (20°C), pointing to irreversible changes in the structure of the material, but the material maintained 20% of the original strength even on heating to 400°C (Fig. 7.2).

120

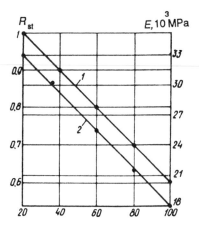

Fig. 7.1. Effect of temperature up to 100°C on strength R_{st} (1) and modulus of elasticity E (2) of FAM polymer concrete prisms.

Table 7.2. Coefficients of reduction of strength and initial modulus of elasticity in FAM polymer concretes depending on temperature

Temperature, °C	Coefficients of reduction of ultimate strength according to the data of		Coefficients of reduction of initial modulus of elasticity according to RCRI and VSEI
	Reinforced Concrete Research Institute (RCRI)	Voronezh Structural Engineering Institute (VSEI)	
20	1	1	1
40	0.9	0.9	0.9
60	0.8	0.8	0.8
80	0.7	0.75	0.7
100	0.6	0.65	0.6

It is known that the strength of thermoplastic polymers, rubbers, and resins increases in direct proportion to the reduction of temperature and the absolute strength of these materials can be realised at −200°C and below.

Very few investigations of the strength characteristics of polymer compositions based on thermosetting resins in relation to the effect of low temperatures have been carried out. These essentially important properties should, therefore, be studied for polymer concretes. Strength characteristics were determined at low temperatures of zero to −195°C on the same specimens as at high temperatures. Cooling to zero and −20°C was carried out in the freezer chambers of the Reinforced Concrete Research Institute and cooling to −40 and −60°C in the

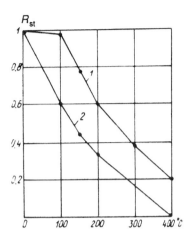

Fig. 7.2. Effect of temperature up to 400°C on strength R_{st} of FAM polymer concrete prisms.

1—after cooling to 20°C; 2—prisms tested in hot condition.

thermal vacuum chamber TBK-1000 and up to -195°C in special thermostats, i.e., containers with liquid nitrogen.

Specimens from the freezer chambers were brought to the testing machines in thermos flasks. To reduce heat transmission from the metal plates of the press during the tests, strips of fibreglass were placed between the working faces of the specimens and the press plates. In the case of polymer concretes based on polyester and epoxide resins, as the temperature dropped an increase in strength up to 30% was noticed at -195°C.

Investigations on polymer concrete specimens based on furan, carbamide, and phenol-formaldehyde resins showed a proportionate increase in strength as the temperature dropped to zero. On reducing the temperature to -20°C, the strength not only did not increase, it diminished and stabilised or increased insignificantly only on further reduction of temperature. Thus, as the temperature dropped no linear increase in strength in the case of these polymer concretes was observed.

The above patterns of the strength variation of polymer concretes can be explained by the fact that these resins contain a definite amount of water. In the temperature interval zero to -20°C, the water present in the pores and capillaries freezes, causing microfissures in the polymer concrete. Therefore, in spite of the increased strength associated with low temperatures, the presence of local microfissures leads to an overall reduction of strength. On further reduction of temperature, although some increase in strength does occur, the effect of the presence of microfissures can be perceived.

7.2. Combustibility of Polymer Concretes

The Reinforced Concrete Research Institute together with the All-Union Research Institute of the Refractory Industry carried out investigations on the inflammability and combustibility of polymer concretes based on different types of binders varying in content from 8 to 15% by weight [66].

Combustibility tests were initially carried out using the method of the All-Union Research Institute of the Refractory Industry in a fire tube. This is an accelerated test providing a preliminary evaluation of the degree of combustibility of the materials. However, judging from the test results (Table 7.3), this method does not take into consideration the specific features of polymer concretes. By this method, a material was assigned to the group of combustible substances if the weight loss of the specimen after the tests exceeded 20% and independent burning with a flame or smoldering continued for over 60 sec. Since the content of the polymer binder in the polymer concrete was less than 20%, on the basis of this method, all types of polymer concretes fall in the group of not readily combustible substances. Another drawback of the fire tube method is the size of the specimen (10 mm × 35 mm × 150 mm). The presence of aggregate grains in polymer concrete calls for a minimum section of not less than 40 mm × 40 mm of the specimens. Thus, specimens for testing in the fire tube may be produced either from polymer slurries, which contain a high content of the polymer binder, or may be cut from much larger polymer concrete intermediate products, which involves certain difficulties.

Table 7.3. Weight loss of specimens after tests in a fire tube

Concrete	Amount of polymer binder, %	Weight loss of specimen, %
Cement grade 400	—	0.15
Polymer concrete based on ED–20 resin	15	7.7
Polymer concrete based on carbamide resin (KF–Zh)	12	0.29
Polymer concrete based on furan resin FAM	8	0.17

Therefore, subsequent combustibility tests were carried out in a ceramic tube (Fig. 7.3) in which specimens 40 mm × 40 mm × 160 mm could be tested. Further, the weight loss after the test was not related to the total weight of the specimen, but to the weight of the polymer binder. Together with the All-Union Research Institute of the Refractory Industry, it was established that polymer concretes can be placed in the group of combustible substances when the weight loss exceeds 9%, and in the group of not readily combustible materials if the loss is below 9%.

After exposure to the effect of fire in the ceramic tube, the prisms were subjected to bending tests and their strength reduction determined and compared

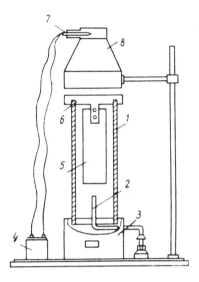

Fig. 7.3. Ceramic tube equipment:

1—ceramic container; 2—gas burner; 3—metal support; 4—automatic potentiometer; 5—specimen; 6—specimen holder; 7—thermocouple; 8—hood.

with the control specimens.

The method of calorimetry has been adopted in the Soviet Union for determining the group of combustible homogeneous solid materials. By calorimetry, the materials are assigned to groups of combustibility, quantitatively based on the index of inflammability K, which is a dimensionless ratio of the quantum of heat liberated by the specimen during the test q_{teo} to the quantum of heat from the ignition source q_n. The classification of materials based on inflammability by calorimetry is shown in Table 7.4. The test results obtained from specimens tested using the ceramic tube (Table 7.5) and the calorimetry method showed that these two methods describe quite objectively the group of inflammability of specimens of various types of polymer concretes. Polymer concretes based on epoxide resins belong to the group of combustible materials. Like polymer concretes based on polyester resins PN–1 and MMA, polymer concretes based on polyester resins PNS–609, PN–62, and PN–63 fall in the group of not readily ignitable substances. Polymer concretes based on carbamide resin KF–Zh and furan resins FAM alone are regarded as not readily combustible.

7.3. Heat Stability of Polymer Concrete Structures

Experience has shown that accelerated tests using a ceramic tube or the calorimetry method can serve only for a preliminary evaluation of the com-

Table 7.4. Inflammability groups according to calorimetry testing

Group	Inflammability index K	Inflammability characteristics
Incombustible	Up to 0.1	Do not burn
Not readily combustible	0.1 to 0.5	Do not support combustion
Not readily ignitable	0.5 to 2.1	Render spread of burning difficult
Combustible	Above 2.1	Burn
Readily ignitable	Above 2.1	Spread burning along horizontal plane

Table 7.5. Results of combustibility tests on polymer concrete specimens in a ceramic tube

Concrete	Weight loss, %		Residual strength after tests, %	Inflammability index K	Group
	total	by weight of polymer binder			
Cement grade 400	2.38	—	100	0.06	Incombustible
Polymer concrete based on epoxide resin ED–20	3.2	9.7	Total loss of strength	—	Combustible
Polymer concrete based on polyester resin PNS–609–22s	0.49	3.3	83	0.53	Not readily ignitable
-do- based on PN–63 resin	0.77	5.1	84	0.47–0.54	-do-
-do- based on KF–Zh resin	0.52	4.33	88.5	0.2	Not readily combustible
-do- based on FAM resin	0.133	1.63	92.7	0.14	Not readily combustible

bustibility of materials. Final data on heat stability of the structures can be obtained only by testing real structures under a standard load.

For fire tests, three columns of section 400 mm × 400 mm and length 3.5 m were prepared. The reinforcement in the columns consisted of four 16 mm diameter longitudinal rods of crimped profile made of steel grade A-II and provided with stirrups of 8 mm diameter made of grade A-I reinforcing bar and spaced 300 mm apart. At the supporting sections, over a length of 320 mm mesh reinforcement of 8 mm diameter having a grid size 70 mm × 70 mm was placed 60 mm apart.

The columns were made of FAM polymer concrete with the following composition, %: FAM furfural-acetone resin 10; benzenesulphonic acid 2; andesite flour 12; silica sand 23; granite aggregates 53; and sodium fluosilicate 1.5 by weight of the resin. Tests on control cubes showed that the strength of the polymer concrete of the three columns averaged 72.5 MPa.

The method of fire testing strove for maximum simulation of the actual working conditions of the structure under fire. Tests were carried out in special furnaces in a temperature regime determined by the standard curve 'temperature—duration of fire'. Simultaneous with heating, conditions of supporting and loading corresponding to their working conditions in installations were created. The columns were hinged at the two ends and tested for compression under a standard axial load applied at the geometric centre of their cross-section.

Furnace temperatures were measured by thermocouples placed on the heated column surface. The thermocouple readings were recorded every 5 min from the beginning to the end of tests.

Visual observations through the inspection window of the furnace during the course of the tests showed identical results of the effect of fire on all the three columns. The ignition of the decomposition products of the polymer concrete on the surface occurred after 4 to 8 min; explosive damage of the surface layer accompanied by slight cracking and formation of funnel-like holes 6 to 8 mm in diameter and 5 to 7 mm in depth commenced after 5–10 min. At some places, large aggregates were exposed and damage continued for 15–25 min without, however, posing any danger to the bearing capacity of the structure as a whole. After 20–35 min of commencing the tests, a coke crust formed on the outer surface and a network of fissures appeared. These fissures opened up further during the course of the tests. Combustion continued along the fissures until the end of the tests. At the end of exposure to the effect of fire, the combustion process of the dissociation products of polymer concrete continued along the fissures for 15–20 min.

The first two columns were tested under a standard load of 130 tonnes, which corresponds to a fourfold long-term safety factor. The heat stability, determined by the duration before the loss of bearing capacity, was practically identical in two of the columns (2 h 7 min and 2 h 6 min). They could be recommended for industrial buildings of the second class of heat stability. The third column tested under the same conditions under a load of 100 tonnes withstood the fire tests for 3 h 2 min. The minimum limit of heat stability of load-bearing structures for buildings of the first class of heat stability is 2.5 hr.

Thus, the results of tests on the effect of fire on load-bearing columns made of FAM polymer concrete and steel confirmed their high heat stability. Such structures can be used for industrial buildings of the first and second classes of heat stability.

By a similar method, tests were carried out at the Order of Lenin and Red Banner of the Labour Moscow Institute of Railway Transport Engineers on the effect of fire on flexural members. Beams (Fig. 7.4) of section 150 mm × 500 mm × 3200 mm were made of FAM polymer concrete according to the following composition, %: furfural-acetone resin FAM 8; benzenesulphonic

126

acid 2; andesite flour 10; silica sand 28; and granite aggregates 52. The average strength of control prisms 70 mm × 70 mm × 280 mm was 70.9 MPa.

Tests on reinforced polymer concrete beams showed that, depending on the thickness of the protective layer (cover concrete) and the percentage of reinforcement, the heat stability ranged from 80 to 135 min. Thus, increasing the polymer concrete protective layer by 10 mm increased the limit of heat stability from 80 to 100 min, i.e., by 25%, while doubling the percentage of reinforcement increased the heat stability to 135 min, i.e., by 1.7 times.

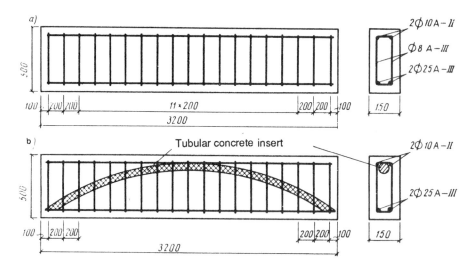

Fig. 7.4. Sketch showing reinforcement of polymer concrete beams:

a—usual reinforcement; b—integrated structure with tubular concrete inserts.

In addition to reinforced polymer concrete beams, integrated beams with rigid tubular concrete arched insert (Fig. 7.4, b) were also tested. However, the use of a rigid tubular insert did not yield the anticipated improvement and the heat stability of such a structure was only 80 min. This comparatively low heat stability of the integrated structure could probably be ascribed to the incorrect selection of the polymer concrete protective layer thickness at 11 mm [31].

Thus, laboratory and field tests on polymer concrete structures helped in understanding the essential features of their breakdown by the effect of temperature and fire.

The strength reduction of conventional cement concretes begins to be perceptibly influenced at 300–400°C or above. At this temperature, irreversible changes begin to occur in the cement matrix as a result of the degradation and decomposition of the hydrosilicates and hydroaluminates of calcium and other newly formed constituents.

In polymer concretes subjected to 100–150°C temperature and in polymer concretes based on polyester resins subjected to 80–100°C, differences in the coefficients of thermal deformations of polymer binder and mineral constituents of polymer concrete are seen and high-elastic and plastic deformations increase significantly. Further, at 150–200°C, the destruction processes of polymer binder begin and a significant amount of hot gaseous products is liberated in the combustion process, and hence the damage of polymer concretes is stimulated due to the exothermal effect of thermo-oxidative destruction of the polymer constituent of the binder.

It should also be pointed out that the low thermal conductivity of polymer binder and the intense gas liberation in the surface layer reduce the heating duration of the deep layers and prevent combustion in the body of the material. As a result, in the outer layer directly exposed to the effect of heat and fire, decomposition occurs initially and the burning of the polymer binder thereafter. Later, the combustion process extends to the deeper layers until complete combustion and loss of strength of the polymer concrete.

The experimental investigations carried out, the proposed method of calculating the heat stability of reinforced polymer concrete beams and the method of calculating the deflections of flexural elements with allowance for the variation of modulus of elasticity and the elastoplastic properties of the reinforcement and the polymer concrete as a result of the action of high temperature, enabled formulation of a more rational design of such structures and designation of the category of their heat stability.

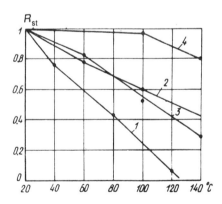

Fig. 7.5. Dependence of strength R_{st} of polymer concrete prisms based on polyester and furan resins on temperature:

1—based on PN–1 resin; 2—PNS–609–21M resin; 3—PN–63 resin; 4—FAM resin.

Polymer concretes based on PN-1 type polyester resins revealed a sharp reduction in strength even with an insignificant rise of temperature (Fig. 7.5).

Specimens of this polymer concrete underwent intense combustion in tests in a ceramic tube. When using polyester resins type PNM–609–21M, PN–63, PN–62, and some others as binder, a less intense reduction of strength was noticed under the action of high temperatures according to the data of Obolduev [53]. In fire tests in a ceramic tube, the specimens practically did not change their weight or external form. Based on these data, polymer concretes based on polyester resins PNS–609–21M, PN–63 and PN–62 can be recommended for use in load-bearing structures in several industrial establishments with a working temperature not exceeding 60°C.

To achieve a more accurate design of polymer concrete structures exposed to the effect of temperature and fire, a method has been proposed [53] for evaluating the thermal stability of polymer concretes in structures, namely, determining the extent of the damaged layer in the first stage of combustion.

In such an approach, an increase in heat and fire stability of polymer concrete structures is achieved by compensating for the inevitable loss of the bearing capacity in the outer portion of the section in the course of combustion by correspondingly increasing the section by increasing the thickness of the layer which may be subjected to destruction.

Moreover, the investigations carried out enabled appropriate improvements in the designing of reinforced polymer concrete structures with allowance for the possible effect of heat during a fire.

8

Design Characteristics of Polymer Concrete and Reinforced Polymer Concrete Structures

Polymer concretes were earlier used predominantly in making products and in the construction of structures without reinforcement: the main requirement then was durability in various aggressive media. The cross-sections of structures were determined by a set of factors, among which strength factors often did not play a decisive role. Under these conditions, satisfactory results were obtained by designing structures for the permissible stresses on the basis of the actual properties of the type of polymer concrete used and fairly high safety factors.

The need for load-bearing structures using reinforced polymer concretes necessitated working out theories and methods of designing reinforced polymer concrete structures. This problem was resolved by Soviet scientists who adopted as a basis the modern method of designing reinforced concrete structures for the limit states and developed it for polymer concrete and reinforced polymer concrete structures intended for working under conditions of exposure to aggressive media and temperatures in the range 80° to −60°C.

For using polymer concretes and reinforced polymer concretes in load-bearing building structures, very exacting requirements of physical, mechanical and chemical properties were prescribed for the polymer concrete mix. These requirements called for taking into consideration not only the effect of external loads and the environment, but also such factors as internal shrinkage and thermal stresses.

Based on statistical processing of the results of innumerable tests on specimens and actual structures, a theory for designing polymer concrete and reinforced polymer concrete structures was proposed by Prof. A.I. Chebanenko [18, 20] and specialists of the Order of Lenin and the Red Banner of the Labour Moscow Institute of Railway Transport Engineers, the State Institute for Designing Non-ferrous Metallurgical Works and the Reinforced Concrete Research Institute developed several standards [35, 38, 39, 40].

The characteristic features of the method for designing reinforced polymer concrete structures and examples of determining the design parameters of polymer concretes based on furan resins and fibreglass reinforcement are given below.

The following are the basic steps in this method: integrated evaluation of mechanical and deformation properties of polymer concretes using stress-strain diagrams whose contours are established based on four characteristics, viz., short-term strength, long-term strength, initial modulus of elasticity and ultimate deformation; consideration of the effect of physical non-linear mechanical characteristics of polymer concretes; evaluation of the rheological properties of the material; and consideration of the characteristic features of prestressed reinforced polymer concrete structures with metal and fibreglass drawn reinforcement.

The design equations are based on the widely recognised method of designing reinforced concrete structures for the limit states of the first and second groups (see below) (Construction Norms and Specifications, SNiP, 2.03.01-84, 'Concrete and Reinforced Concrete Structures').

Further, it should be remembered that the load-bearing capacity and reliability of polymer concrete structures are directly associated not only with the type of polymer concrete selected, but also the working conditions. All these factors should be taken into consideration to the extent necessary when designing each given structure.

8.1. Polymer Concretes

In producing polymer concrete and reinforced polymer concrete structures, heavy and light-weight polymer concretes with dense structure and compressive strength 30 to 90 MPa are used. For prestressed structures, polymer concretes containing not more than 10% synthetic binder are recommended to reduce the loss of prestressing due to shrinkage and creep.

Average Indexes of Basic Physical Properties of Structural
Polymer Concretes Based on FAM Monomer

Brinell hardness, MPa	4.1
Poisson's ratio	
for heavy polymer concrete	0.25
for polymer concrete based on porous aggregates	0.21
Abradability against a loosely fixed abrasive, %	40
Thermal conductivity, kJ/(hr-m-°C)	$(0.6-0.75) \times 4.1868$
Volume shrinkage, %	0.3
Linear shrinkage, %	0.1
Water absorption after 24 hr, %	0.17
Heat stability, °C	170
Cold resistance, cycles, above	300
Thermal coefficient of linear expansion, 1/°C	$(12-20) \, 10^{-6}$

The main mechanical characteristics of polymer concrete are the magnitude of axial compressive strength B and the magnitude of axial tensile strength B_t, which should correspond to the guaranteed strength values of the polymer concrete with a reliability of 0.95. These characteristics are determined from the data of short-term tests using standard specimens (for compression, a standard cube specimen 15 cm $\times$ 15 cm $\times$ 15 cm) at normal temperature and humidity conditions.

The standard short and long-term strength parameters of polymer concrete are the axial compressive strengths of the prism (prism strength) R_{bn} and R'_{bn} and axial tensile strengths R_{btn} and R'_{btn}. These values are established using the following equations:

$$R_{bn} = [(1 - 1.64\nu_b)/1.25]\overline{B} , \qquad \ldots (8.1)$$
$$R_{btn} = [(1 - 1.64\nu_{bt})1.2]\overline{B}_t , \qquad \ldots (8.2)$$

where $\overline{B}$ and $\overline{B}_t$ are the average values of the strength of standard polymer concrete specimens under compression and tension; ν_b and ν_{bt} the variation coefficients of the compressive and tensile strengths respectively of the polymer concrete.

The standard long-term strengths corresponding to the action of permanent and long-term loads are established using the following equations:

$$R'_{bn} = R_{bn}(1 - E_b\epsilon_{bc}^{max}/R_{bn}) , \qquad \ldots (8.3)$$
$$R'_{btn} = R_{btn}(1 - E_b\epsilon_{btc}^{max}/R_{btn}) , \qquad \ldots (8.4)$$

where ϵ_{bc}^{max} and ϵ_{btc}^{max} are the maximum average values of relative deformations of the creep of polymer concrete under compression and tension respectively, which are reversible in time (elastic deformations); under compression, ϵ_{bc}^{max} is roughly 15 to 20% ϵ'_b; and E_b the initial modulus of elasticity determined with reliability 0.95 using equation

$$E_b = \overline{E}_b(1 - 1.64\nu_b) , \qquad \ldots (8.5)$$

where, $\overline{E}_b$ is the mean arithmetical value of the modulus of elasticity of polymer concrete determined at $\sigma_b \leq 0.3 R_{bn}$ in the short-term loading regime of standard specimens at a rate of not less than 60 MPa/min.

In equation (8.1), a value of 1.25 characterises the changeover from cube to prism (cylinder) strength of the polymer concrete while a coefficient of 1.2 in equation (8.2) guarantees the least level of the stressed state of the material at which its destruction by short-term loading under tension is least probable or eliminated altogether.

The theoretical short-term and long-term strengths of polymer concrete for the limit states for the first group under compression and tension, namely, R_b and R_{bt} and R'_b and R'_{bt}, are determined by dividing their standard values by the safety factor adopted, i.e., 1.3.

The values of maximum relative deformations corresponding to the theoretical long-term strengths are usually regarded as invariable when calculating for the second as well as first groups of limit states, i.e., $\epsilon'_b = \epsilon_{bn}$ and $\epsilon'_{bt} = \epsilon_{btn}$.

The main theoretical characteristics of polymer concretes based on furan resins are shown in Table 8.1.

Standard and theoretical long-term strengths of polymer concrete under compression and tension are generally determined by multiplying the corresponding long-term strengths R'_b, R'_{bt} and R'_{bn}, R'_{btn} (Table 8.1) by the coefficient j:

$$j = P \left/ \left(P_{lt} + \frac{R'}{R} P_{st} \right) \right. , \qquad \ldots (8.6)$$

where R' is the long-term strength of polymer concrete under compression or tension (theoretical or standard); R the short-term strength of polymer concrete under compression or tension (theoretical or standard); P_{st}, P_{lt}, and P the forces due to short-term, long-term (including permanent) and total loads applied at load coefficient $n > 1$ or $n = 1$.

When determining the long-acting (sustained) loads from among the temporary loads, the long-acting portion of the load will have to be isolated; for snow loads, it is 80%, for overhead loads and loads with an active duration of one to 10 days 70%, and wind loads 30%. When designing sections of eccentrically loaded members, the corresponding moments relative to the axis traversing through the centre of gravity of tension or compression zone are taken as forces P_{st}, P_{lt}, and P. In the absence of short-term loads $(P_{st} = 0)$, the standard and theoretical strengths are correspondingly equal to the long-term strengths (Table 8.1).

Coefficient j in equation (8.6) should fall in this range: under compression—

$$R_{bn}/R'_{bn} = R_b/R'_b \geq j \geq 1 , \qquad \ldots (8.7)$$

under tension—

$$R_{btn}/R'_{btn} = R_{bt}/R'_{bt} \geq j \geq 1 . \qquad \ldots (8.8)$$

The strength of polymer concrete $(R_b, R_{bt}, R_{bn}, R_{btn})$ introduced in the calculation is generally determined using the equation:

$$R = jR' \overset{i}{\underset{1}{n}} \gamma_{bi} , \qquad \ldots (8.9)$$

where γ_{bi} is the product of the coefficients of the working conditions of the material, established from Tables 8.2, 8.3 and 8.4.

Notes. 1. Theoretical strengths of polymer concrete for the second group of limit states are taken as equal to the standard strengths and are introduced in the calculation with a single coefficient of the working conditions taken as the least of the total number of limit states considered when calculating the first group of limit states.

2. When designing the structures for strength in the stage of preliminary compression or during transport and assembly, the theoretical strengths of polymer concrete are multiplied by coefficient $\gamma_b = 1.15$.

Table 8.1. Main theoretical characteristics of polymer concretes based on furan resins

Index	Symbol	Heavy polymer concretes		Fine-grained		Light polymer concretes	
		FAM-8	FAM-10	FAM-M1	FAM-M2	Agloporite FAM-A	Claydite FAM-K
Density, kg/m^3	ρ	2300	2300	2200	2200	1750	1600
Magnitude of compressive strength, MPa	B	60	60	70	80	50	25
Standard short-term strength, MPa, under							
axial compression	R_{bn}	45	45	52	58	39	19.5
axial tension	R_{btn}	3.9	3.9	5.2	6.5	3.9	2.3
Standard long-term strength, MPa, under							
axial compression	R'_{bn}	27	27	31	35	23	11.5
axial tension	R'_{btn}	1.3	1.3	1.7	2.1	1.3	0.76
Theoretical short-term strength, MPa, under							
axial compression	R_b	34	34	40	44	30	15
axial tension	R_{bt}	3	3	4	5	3	1.8
Theoretical long-term strength, MPa, under							
axial compression	R'_b	20	20	24	27	17.5	8.5
axial tension	R'_{bt}	1	1	1.3	1.6	1	0.58
Initial modulus of elasticity, MPa	E_b	3×10^4	2×10^4	2×10^4	2.5×10^4	1.8×10^4	1.5×10^4
Theoretical long-term deformation under							
axial compression	ϵ'_b	2.8×10^{-3}	4.2×10^{-3}	4.8×10^{-3}	4.3×10^{-3}	4.1×10^{-3}	2.4×10^{-3}
axial tension	ϵ'_{bt}	3.7×10^{-4}	5.6×10^{-4}	6.4×10^{-4}	5.7×10^{-4}	5.5×10^{-4}	3.2×10^{-4}

Table 8.2. Coefficient of working conditions γ_{b1}

Temperature of the medium, °C	Compression	Tension
20	1	1
40	0.9	0.8
60	0.8	0.7
80	0.7	0.6

Note: Coefficient γ_{b1} for other temperatures of the medium is determined by interpolation.

Table 8.3. Coefficient of working conditions γ_{b2}

Humidity characteristics of environment	Compression	Tension
Relative humidity, %:		
Constant	—	—
60	1	1
80	0.75	0.7
97	0.7	0.6
Variable 50–90%	0.7	0.6
Submergence in water	0.6	0.4

Note: When using ground andesite instead of ground quartz sand as a microfiller, the coefficient of working conditions γ_{b2} is raised by 15%.

The relative deformation ϵ under compression and tension introduced in the calculation is established according to the following equation:

$$\epsilon = \frac{1 - j(R'/R)}{1 - R'/R}\left(\epsilon'_b - \frac{R}{E_b}\right) + \frac{R}{E_b}, \qquad \ldots (8.10)$$

where ϵ'_b is the maximum theoretical deformation under compression ϵ'_b or tension ϵ'_{bt}.

The theoretical deformations of the polymer concrete under compression $\epsilon_b, \epsilon_{bn}$ and under tension $\epsilon_{bt}, \epsilon_{btn}$ are raised compared to their values established using equation (8.10) by dividing by the coefficient of working conditions which is the least of those taken into consideration in equation (8.9). Further, in all cases, the value of ϵ should satisfy the condition

$$\epsilon \leq \epsilon'. \qquad \ldots (8.11)$$

Note. Under the combined action of temperature, chemical reagents and moisture, the theoretical deformation ϵ determined using equation (8.10) should be divided only by the coefficient of working conditions of the material γ_{b1} which takes into consideration the effect of high temperature.

Table 8.4. Coefficient of working conditions γ_{b3}

Reagent	Maximum concentration, %	Coefficient
Mineral acids:		
sulphuric	70	1
	85	0.9
hydrochloric	40	1
phosphoric	5	0.75
Organic acids:		
lactic, oleic	85	1
citric	10	1
acetic	5	1
Salts and bases:		
ammonia	25	1
caustic soda	50	1
sodium carbonate	20	1
Chloride solutions of salts and metals		
(iron, potassium, calcium, magnesium,		
zinc)	100	1
Solvents		
(aniline, benzene, alcohols, turpentine,		
toluene, phenol)	100	1
Gases		
(chlorine, carbon dioxide, hydrogen		
sulphide, carbon disulphide, hydrogen		
chloride)	100	1
Formaldehyde	50	1

Note: For other concentrations, coefficient γ_{b3} is established experimentally.

The shear modulus of polymer concrete G is taken as equal to 0.25 of the corresponding values of theoretical modulus of elasticity E_b.

Furfural-acetone polymer concretes are stable to all acids (with the exception of oxidising acids, such as nitric, chromic, etc. of a concentration exceeding 10%), alkalis (with the exception of hypochlorides) and salt solutions; they are moderately stable in water but unstable in acetone (see Table 8.4).

8.2. Reinforcement

For reinforcing polymer concrete structures, all types of steel reinforcement except thermally and thermo-mechanically hardened steel as also fibreglass are used. The latter is generally used as prestressed reinforcement. The diameter of the reinforcing rods should be not less than 4 mm including the wires used for making the reinforcing cables.

The standard and theoretical characteristics of steel reinforcements are adopted in accordance with Construction Norms and Specifications, SNiP, 2.03.01-84 with the following supplementary conditions.

Non-prestressed reinforcement placed in the section of a member compressed by external loads (reinforcement area A') is introduced in the calculation with theoretical strength R_{sc} taken as equal to R_s but not more than $\epsilon_s E_s$ and not more than 600 MPa. Here, the theoretical relative deformation ϵ_s is determined using equation (8.10).

The theoretical strengths of steel reinforcement for limit states of the first group are reduced (or raised) by multiplying by the corresponding coefficients of working conditions γ_s according to the Construction Norms and Specifications, SNiP, 2.03.01-84. The coefficient of working conditions γ_{s6} (item 6, Table 24 of SNiP) is taken as equal to one when designing reinforced polymer concrete structures.

The length of transmission zone I_p for prestressed steel reinforcement without anchors is determined using the following equations:

for reinforcing rods with a crimped profile

$$I_p = [0.6(\sigma_{sp}/R_{cp}) + 20]d , \qquad \ldots (8.12)$$

for high strength wire grade Vr-II and reinforcing cables K-7 and K-19:

$$I_p = [3.6(\sigma_{sp}/R_{cp}) + 100]d , \qquad \ldots (8.13)$$

where σ_{sp} when designing members for strength is taken as equal to R_s and σ_{sp}, whichever is greater; when designing members for fracture resistance, it is taken as equal to σ_{sp}. Here, σ_{sp} is adopted with allowance for the initial prestress losses from Table 8.5; R_{cp} is the transfer compressive strength of the concrete.

All along the length I_p, meshes or rings are placed with the distance between the transverse rods at not more than $10d$ (d = diameter of the prestressed reinforcement in cm) and not less than 5 cm.

The anchor zone length I_s of the non-prestressed steel reinforcement in the polymer concrete under tension is determined using the following equations:

for rod reinforcement with crimped profile

$$I_s = [1.2(R_s/R_{bn}) + 1.6]d , \qquad \ldots (8.14)$$

for smooth reinforcing rods

$$I_s = [1.5(R_s/R_{bn}) + 16]d . \qquad \ldots (8.15)$$

Further, the smooth reinforcing rods should end in hooks or have welded cross reinforcement over length I_s. The cross rods (rings) or meshes are distributed all along length I_s and placed such that the distance between the cross rods is not more than 10 d (where d is the diameter of the longitudinal reinforcement) and not less than 8 cm.

Characteristics of Fibreglass Reinforcement of Crimped Profile and Diameter 6 mm used in the USSR

Density ρ, kg/m^3	1900
Mean arithmetical value of short-term fracture, $\overline{R}_s$, MPa	1450
Standard tensile strength (with reliability 0.98), MPa:	
short-term R_{sn}	1250
long-term R'_{sn}	810
Modulus of elasticity E_s in loaded section (0.2 to 0.5) R_{sn}, MPa	50,000
Theoretical tensile strength, MPa:	
short-term R_s	800
long-term R'_s	520
Coefficient of linear expansion (aluminoborosilicate fibre reinforcement) α_s, 1/$^\circ$C	5.8 $\times$ 10^{-6}

Note. Mechanical characteristics pertain only to fibreglass reinforcement under tension and correspond to the standard temperature and moisture conditions of the environment.

The fibreglass reinforcement is prestressed only mechanically. Sudden application of compressive force on polymer concrete for fibreglass reinforcement is not permitted.

The tensile strength of fibreglass reinforcement (theoretical R_s and standard R_{sn}) is determined by multiplying the corresponding long-term tensile strengths R'_s and R'_{sn} by coefficient j:

$$j = P \left/ \left(P_{lt} + \frac{R'_s}{R_s} P_{st} \right) \right. , \qquad \ldots (8.16)$$

where R'_s is the long-term tensile strength of the fibreglass reinforcement, theoretical or standard (R'_s, R'_{sn}); and R_s the short-term tensile strength of fibreglass reinforcement, theoretical or standard (R_s, R_{sn}).

Further, the strengths R'_s, R_s and R'_{sn}, R_{sn} corresponding to forces P_{lt}, P_{st}, and P (with loading coefficient $n > 1$) and P_{lt}, P_{st}, and P (with loading coefficient $n = 1$) are used.

When there are no short-term loads, R_s and R'_s are taken as equal to R'_s and R'_{sn}. The strength values of fibreglass reinforcement for equation (8.16) are taken without change. The values of theoretical tensile strength R_s in the calculation for the first group of limit states for fibreglass reinforcement are multiplied by coefficients of working conditions γ_s, taking into consideration the effect of high temperatures γ_{st}, aggressive medium γ_{sa}, and also the reduction of cohesion on heating the product γ_{sc}.

Table 8.5. Losses of prestressing, MPa

Factors causing losses of prestressing of reinforcement	Stress in the reinforcement (according to SNiP 2.03.01-84)	
	on supports	in polymer concrete
A. Initial Losses		
Relaxation of stresses in reinforcement σ_1	Item 1, Table 5	—
Exothermal self-heating, heating, shrinkage, and rapid creep σ_2	100—at open surface of member to its volume ratio over 15 m^{-1}; 120—less than 15 m^{-1}	—
Deformation of anchors placed at the tensioning devices σ_3	Item 3, Table 5	Item 3, Table 5
Friction of reinforcement against channel walls, surface of polymer concrete structure and bending device σ_4	Item 4, Table 5	Item 4, Table 5
Deformation of steel mould when producing prestressed reinforced polymer concrete structure σ_5	Item 5, Table 5	—
Shrinkage and rapid creep of polymer concrete σ_6	—	50—at member surface face to its volume ratio over 15 m^{-1}; 60—less than 15 m^{-1}
B. Secondary Losses		
Relaxation of stresses σ_7	—	Item 7, Table 5
Creep of polymer concrete σ_8	—	—
Crumpling under threads of spiral or circular reinforcement (at diameter of member up to 5 m) σ_9	—	50
Compressive deformation of joints between blocks σ_{10}	—	Item 11, Table 5
Aggressive effect of medium σ_{11}	—	—

The coefficients of working conditions for fibreglass reinforcement are given in Table 8.6.

The simultaneous use of steel and fibreglass reinforcement is permissible and both of them or only the fibreglass reinforcement can be prestressed. If a prestressed reinforcement forms the base of the element, ensuring its working conditions in both the groups of limit states, the non-prestressed reinforcement should attain only such stresses which correspond to the more rational use of the prestressed reinforcement.

The use of fibreglass reinforcement in reinforced polymer concrete structures subjected to the action of dynamic or pulsating loads calls for special justification and experimental testing.

Table 8.6. Coefficient of working conditions for fibreglass reinforcement

Factors necessitating the use of coefficients	Symbol	Index
Effect of high temperatures:		
short-term heating in dry state at not over 2°C/min to 80°C	γ_{st-1}	0.95
long-term effect of temperature 80°C heating of FAM	γ_{st-2}	0.9
polymer concrete at 5°C/min to 30°C	γ_{st-3}	0.65
cohesion strength on heating to 80°C	γ_{st-4}	0.8
Effect of aggressive media on structure (during use):		
1 N H_2SO_4 solution	γ_{sa-1}	0.7
salt solutions	γ_{sa-2}	0.8
1 N NaOH solution	γ_{sa-3}	0.8
Effect of water on structure	γ_{sa-4}	0.8

Note: Fibreglass reinforcement based on alkali-resistant fibreglass is recommended when the structure is exposed to acid solutions.

The theoretical strength and the modulus of elasticity of the fibreglass reinforcement for the limit states of the second group are taken as equal to the standard strength R_{sn} and modulus E_s and are used in the calculations with the coefficient of working conditions of the fibreglass reinforcement, the least of the total number considered, when calculating for the first group of limit states.

The use of fibreglass reinforcement in FAM reinforced polymer concrete structures with acidic hardener working under humid conditions is equated to working in an acidic medium.

Under simultaneous action of various intensely aggressive factors, the coefficients of working conditions of fibreglass reinforcement should be established from the test results.

The zone length of stress transmission I_p for the prestressed fibreglass reinforcement is determined using the following equation:

$$I_p = 1.2[3.6(\sigma_{sp}/R_{cp}) + 100]d , \qquad \ldots(8.17)$$

where σ_{sp} is taken as equal to the higher of R_{sn} and σ_{sp} when calculating the strength of the members; and R_{sn} when calculating the fracture resistance of the members.

Within the zone of stress transmission I_p, loops or rings of steel reinforcement are placed with distance not more than 10 d between the transverse rods (d is the diameter of the prestressed fibreglass reinforcement) but not less than 5 cm.

8.3. Additional Requirements for Designing Reinforced Polymer Concrete Structures

Some basic prerequisites. Before designing an actual structure, the values (based on control tests) of shrinkage and thermal deformations in polymer concrete should be determined and the corresponding internal stresses calculated using the following equations:

$$\sigma_y = \frac{y_{max}E_b A\psi_1}{(1-\mu)(1+m)10^3} , \qquad \ldots(8.18)$$

where y_{max} is the maximum shrinkage (relative) of polymer concrete; A the elastic relative deformation, $A = 0.3\ldots0.4$; μ the Poisson's ratio; and ψ_1 the relaxation coefficient of shrinkage stresses, $\psi_1 = 0.8\ldots0.9$;

$$m = E_b A_{red}/E_s A_s = (E_b/E_s)(1/\mu)10^2 . \qquad \ldots(8.19)$$

The following condition should also be satisfied:

$$\sigma_y \leq R_{btn} . \qquad \ldots(8.20)$$

If condition (8.20) is not satisfied, the composition of the polymer concrete should be reviewed and the amount of the binder reduced.

$$\sigma_t = \alpha_b(t_c - t_n)E_b K\psi/(1-\mu) , \qquad \ldots(8.21)$$

where σ_t is the thermal stress in the polymer concrete; K, a proportionality coefficient (0.95); t_c and t_n the self-heating temperatures at the centre and on the surface of the structure, °C (at ambient temperature 20°C, $t_n = 30\ldots40$°C and $t_c = 50\ldots80$°C, minimum values for thin-walled structures); μ the Poisson's ratio, $\mu = 0.34\ldots0.4$ at $t = 60$ to 80°C; and ψ the coefficient of relaxation of shrinkage stresses, $\psi = 0.6$ to 0.7.

The following condition should be satisfied:

$$\sigma_t \leq R_{btn} . \qquad \ldots(8.22)$$

If condition (8.22) is not satisfied, the composition of the polymer concrete should be reviewed and the amount of the binder reduced or the geometric shape of the structure modified such that the heat transfer with the ambient atmosphere improves.

When using polymer concrete structures under conditions of constant contact with an aggressive liquid, diffusive mass transfer should be taken into consideration. This should be used when determining the thickness of the protective layer of the reinforcement in the form of a tentative function of the depth of penetration according to equation:

$$h_t = h_\infty[1 - \exp(ZDt)] , \qquad \ldots(8.23)$$

where h_∞ is the depth of penetration in time t, days, $t \to \infty$; Z a constant for the sample, $Z = \pi \times 3600 \times 24/\delta^2$ (here, δ is the thickness of the specimen, cm); D the diffusion coefficient, cm²/sec; for ordinary compositions $D = (6 \text{ to } 8) \times 10^{-8}$ cm²/sec and for compositions with carbon-containing fillers and aggregates $D = (1.5 \text{ to } 2) \times 10^{-8}$ cm²/sec.

In equation (8.23), $h_\infty = 5$ cm.

The duration in years over which the aggressive medium passes through a protective layer of reinforcement of thickness $h_t \leq 4$ cm is determined by resolving equation (8.24):

$$t = 51 \ln[5/(5 - h_t)] \ . \qquad \qquad \ldots (8.24)$$

Depending on the degree of aggressiveness of the medium and the type of reinforcement, the recommended width of short-term opening up of normal cracks in reinforced polymer concrete structure varies from 0.05 to 0.3 mm. In such structures, the development of transverse cracks as well as normal cracks near their support sections and in the most compressed zones of structures subjected to bending is not permissible.

8.4. Recommendations for Design

Polymer concrete structures should be of simple design allowing for the possibility of their production in single-piece moulds. During production, such structures are usually subjected to heat treatment. Further, the drop-test (impact) strength of polymer concrete structures should be not less than 90% of the designed strength. The striking (demoulding) strength of polymer concrete in structures not subjected to heat treatment should be not less than 70% and the drop-test strength not less than 80% of the designed strength. In prestressed structures, the direct compressive strength of hardened polymer concrete at the time of transfer of prestress should be not less than 80% of the designed strength. After tightening (post-tensioning) the reinforcement in the hardened polymer concrete, the ducts used for inserting the reinforcing members are to be filled with the polymer slurry.

The dimensions of polymer concrete and reinforced polymer concrete sections should allow for the effect of aggressive factors, economic requirements, and their production technology. The elasticity of eccentrically compressed members in any direction should generally not exceed 150 and for columns 100. The thickness of assembled and solid reinforced polymer concrete panels should be not less than 50 mm. The maximum dimensions of sections of polymer concrete structures are determined by the following conditions: when the polymer concrete mix is laid in many layers, the interval between the first and the subsequent layers should not exceed the interval after which intense hardening of the mix commences; the ratio of the exposed surface of the structure to its volume should be not less than 15 m^{-1} at which the self-heating temperature of solid polymer concrete does not exceed 80°C.

The thickness of the protective layer (cover) of polymer concrete in structures used under highly aggressive conditions should be based on calculations for diffusion penetration or established from experimental data.

The thickness of the protective layer for a longitudinal steel reinforcement should be not less than 20 mm in panels and walls of thickness up to 120 mm, and 30 mm in columns, beams and edges, panels and walls of thickness over 120 mm. The thickness of the protective layer for other reinforcements is taken in the range 15–20 mm.

In sections transmitting the force of prestressed reinforcement to the polymer concrete, the thickness of the protective layer should be not less than 50 mm for rod reinforcement of all grades and not less than 30 mm for reinforcing cables and fibreglass reinforcement. The ends of the stressed steel reinforcement and also the anchors should be protected by a thick layer of polymer slurry to a thickness not less than 20 mm; direct positioning of anchors on the surface is not permissible.

The thickness of the protective polymer concrete layer in foundation beams and foundation assemblies should be not less than 40 mm; in solid foundations with concrete preparation, it should be not less than 50 mm and when there is no such preparation, 70 mm.

In flexural, eccentrically compressed, and eccentrically strained members of reinforced polymer concrete structures with a bi-axial stress diagram, the placement of reinforcement in several rows along the height of the section is not recommended.

In reinforced polymer concrete structures with inserted, prestressed steel reinforcement, anchoring of the reinforcement is done by a single method or a combination of methods (Handbook for Designing Prestressed Reinforced Concrete Structures Made of Heavy Concrete).

When using inserted reinforcement in the form of high-strength wires of crimped profile, reinforcing cables K-7 and K-19 and reinforcing rods with crimped profile tightened at the supports, the placement of permanent anchors at the ends of members is not generally required if the stationary temperature regime of the ambient atmosphere does not exceed 50°C.

The placement of permanent anchors at the ends of inserted steel reinforcement is obligatory: when using structures in a medium with a stationary temperature exceeding 50°C; for reinforcements inserted in polymer concrete; when the cohesion of the reinforcement with the polymer concrete is inadequate and when the fabrication measures do not guarantee the absence of cracks in the section of stress transmission.

8.5. Determining the Design Parameters of Polymer Concrete and Fibreglass Reinforcement

Example 1. Given a column of section 40 cm × 50 cm made of FAM-10 polymer concrete with strength characteristics as shown in Table 8.1 and symmetric reinforcement grade A-III. The longitudinal forces and the bending moments due to permanent and long-term loads are $P_{lt} = 1800$ kN, $P_{lt}^n = 1565$ kN; and due to short-term loads $P_{st} = 250$ kN and $P_{st}^n = 217$ kN.

The atmospheric temperature is $+20°C$, relative humidity (variable) 50 to 90%, and the degree of aggressiveness of the medium moderate.

The coefficients which take into consideration the duration of the effect of load can be determined using equation (8.6): under compression:

$$j_b = P \left/ \left(P_{lt} + \frac{R'_b}{R_b} P_{st} \right) \right. = 2050 \left/ \left(1800 + \frac{20}{34} 250 \right) \right. = 1.05 \ ;$$

$$j_{bn} = P^n \left/ \left(P^n_{lt} + \frac{R'_{bn}}{R_{bn}} P^n_{st} \right) \right. = 1782 \left/ \left(1565 + \frac{27}{45} 217 \right) \right. = 1.05 \ ;$$

under tension:

$$j_{bt} = P \left/ \left(P_{lt} + \frac{R'_{bt}}{R_{bt}} P_{st} \right) \right. = 2050 \left/ \left(1800 + \frac{1}{3} 250 \right) \right. = 1.09 \ ;$$

$$j_{btn} = P^n \left/ \left(P^n_{lt} + \frac{R'_{btn}}{R_{btn}} P^n_{st} \right) \right. = 1782 \left/ \left(1565 + \frac{1.3}{3.9} 217 \right) \right. = 1.09 \ .$$

The coefficients of the working conditions of polymer concrete can be taken from Tables 8.2, 8.3, and 8.4: under compression:

$$\gamma_{b1} = 1; \quad \gamma_{b2} = 0.7; \quad \gamma_{b3} = 1 \ ;$$

under tension:

$$\gamma_{b1} = 1; \quad \gamma_{b2} = 0.6; \quad \gamma_{b3} = 1 \ .$$

The short-term and long-term strengths of the polymer concrete used in the calculation can be determined from equation (8.9):

$$R'_b = j_b R'_b n \gamma b_i = 1.05 \cdot 20 \cdot 0.7 = 14.7 \ \text{MPa} \ ;$$

$$R_b = j_b R_b n \gamma b_i = 1.05 \cdot 27 \cdot 0.7 = 19.8 \ \text{MPa} \ ;$$

$$R'_{bt} = j_{bt} R'_{bt} n \gamma b_i = 1.09 \cdot 1 \cdot 0.6 = 0.65 \ \text{MPa} \ ;$$

$$R_{bt} = j_{bt} R_{bt} n \gamma b_i = 1.09 \cdot 1.3 \cdot 0.6 = 0.85 \ \text{MPa} \ .$$

The values of relative deformations under compression and tension used in the calculation can be determined using equation (8.10):

$$\epsilon_b = \frac{1 - j_b(R'_b/R_b)}{1 - R'_b/R_b} \left(\epsilon'_b - \frac{R_b}{E_b} \right) + \frac{R_b}{E_b} = \frac{1 - 1.05 \cdot 20/34}{1 - 20/34}$$
$$\left(4.2 \cdot 10^{-3} - \frac{34}{2 \cdot 10^4} \right) + \frac{34}{2 \cdot 10^4} = 4.02 \cdot 10^{-3} \ ;$$

$$\epsilon_{bn} = \frac{1 - 1.05 \cdot (27/45)}{1 - 27/45} \left(4.2 \cdot 10^{-3} - \frac{45}{2 \cdot 10^4} \right) + \frac{45}{2 \cdot 10^4} = 4.05 \cdot 10^{-3} \ ;$$

$$\epsilon_{bt} = \frac{1 - 1.09(1/3)}{1 - 1/3} \left(5.6 \cdot 10^{-4} - \frac{3}{2 \cdot 10^4} \right) + \frac{3}{2 \cdot 10^4} = 5.4 \cdot 10^{-4} \ ;$$

$$\epsilon_{btn} = \frac{1 - 1.09(1.3/3.9)}{1 - 1.3/3.9} \left(5.6 \cdot 10^{-4} - \frac{3.9}{2 \cdot 10^4} \right) + \frac{3.9}{2 \cdot 10^4} = 5.4 \cdot 10^{-4} \ .$$

The calculated deformations of polymer concrete are raised by dividing by the coefficient of working conditions representing the least of those under consideration; further, in all cases, condition (8.11) should be satisfied, i.e. $\epsilon \le \epsilon'$;

$$\epsilon_b = 4.02 \cdot 10^{-3}/0.7 = 5.75 \cdot 10^{-3} > \epsilon_b' = 4.2 \cdot 10^{-3} \ ;$$
$$\epsilon_{bn} = 4.05 \cdot 10^{-3}/0.7 = 5.8 \cdot 10^{-3} > \epsilon_b' = 4.2 \cdot 10^{-3} \ ;$$
$$\epsilon_{bt} = \epsilon_{btn} = 5.4 \cdot 10^{-4}/0.6 = 9 \cdot 10^{-4} > \epsilon_{bt}' = 5.6 \cdot 10^{-3} \ .$$

For calculation, the following values are adopted:

$$\epsilon_b = \epsilon_{bn} = 4.2 \times 10^{-3}; \quad \epsilon_{bt} = \epsilon_{btn} = 5.6 \times 10^{-3} \ .$$

Example 2. Given a reinforced polymer concrete prestressed ribbed panel with fibreglass reinforcement of diameter 6 mm. The characteristics of FAM-10 polymer concrete are given in Table 8.1 and of fibreglass reinforcement in Table 8.6.

The temperature of the atmosphere is $+20°C$; relative humidity constant at 80%; degree of aggressiveness of the medium moderate; and chlorine concentration goes up to 30%. The load on the panel is as follows: long-term $N_{lt} = 5290 \ N/m^2$, short-term $P_{st} = 280 \ N/m^2$, total $P = 5570 \ N/m^2$ ($P_{lt}^n = 4600 \ N/m^2$, $P_{st}^n = 240 \ N/m^2$, and $P^n = 4840 \ N/m^2$).

The coefficients allowing for the duration of the effect of load on the polymer concrete are determined using equation (8.6):

under compression:

$$j_b = \frac{5570}{5290 + (20/34)280} = 1.02 \ ;$$

$$j_{bn} = \frac{4840}{4600 + (27/45)240} = 1.02 \ ;$$

under tension:

$$j_{bt} = \frac{5570}{5290 + (1/3)280} = 1.03 \ ;$$

$$j_{btn} = \frac{4840}{4600 + (1.3/3.9)240} = 1.03 \ .$$

The coefficients of working conditions of polymer concrete are taken from Tables 8.2, 8.3, and 8.4:

under compression: $\gamma_{b1} = 1; \quad \gamma_{b2} = 0.75; \quad \gamma_{b3} = 1;$
under tension: $\gamma_{b1} = 1; \quad \gamma_{b2} = 0.7; \quad \gamma_{b3} = 1.$

The strength of polymer concrete used in the calculation is determined using equation (8.9):

$$R_b' = 1.02 \cdot 20 \cdot 0.75 = 15.3 \ MPa \ ;$$
$$R_b = 1.02 \cdot 27 \cdot 0.75 = 20.7 \ MPa \ ;$$
$$R_{bt}' = 1.03 \cdot 1.0 \cdot 7 = 0.72 \ MPa \ ;$$
$$R_{bt} = 1.03 \cdot 1.3 \cdot 0.7 = 0.94 \ MPa \ .$$

The values of relative deformations under compression and tension are determined using equation (8.11). Since coefficient j are roughly equal to one,

$$\epsilon_b = \epsilon_{bn} = \epsilon'_b = 4.2 \cdot 10^{-3} \; ;$$
$$\epsilon_{bt} = \epsilon_{btn} = \epsilon'_{bt} = 5.6 \cdot 10^{-4} \; .$$

The coefficients allowing for the duration of the action of load on the fibreglass reinforcement are determined using equation (8.16):

$$j_s = P \left/ \left(P_{lt} + \frac{R'_s}{R_s} P_{st} \right) \right. = 5570/[5290 + (520/800)280] = 1.02 \; ;$$

$$j_{sn} = P^n \left/ \left(P^n_{lt} + \frac{R'_{sn}}{R_{sn}} P^n_{st} \right) \right. = 4840[4600 + (810/1250)240] = 1.02 \; .$$

The coefficients of working conditions of the fibreglass reinforcement are taken from Table 8.6, $\gamma_{sa-2} = 0.8$. The strengths of the fibreglass reinforcement under tension used in the calculation are as follows:

$$R'_s = j_s R'_s \gamma_{sa2} = 1.02 \cdot 520 \cdot 0.8 = 424 \text{ MPa} \; ;$$
$$R_s = j_s R_s \gamma_{sa2} = 1.02 \cdot 810 \cdot 0.8 = 660 \text{ MPa};$$
$$R_{sn} = j_{sn} R_{sn} \gamma_{sa2} = 1.02 \cdot 1250 \cdot 0.8 = 1020 \text{ MPa} \; .$$

9

Technology of Producing Polymer Concrete Products

9.1. Specifications for Process Equipment for Producing Polymer Concrete Products and Structures (FRG)

The production method is not the least of the factors that determine the properties of polymer concretes. The problem of producing high-quality materials possessing optimum characteristics calls for optimum methods of their production.

Two essentially different methods of producing polymer concretes are recognised: batch and continuous methods. Both methods have distinct advantages depending on the objectives set out, requirements skills of workers, planned production indexes, and so forth.

The main link in any technological flow sheet determining in particular the technological process and the period required for producing the materials and products is the mixer unit in which the polymer concrete mix is produced. Various constituents of the reaction are mixed in the mixer and the process once initiated cannot thereafter be stopped. It is therefore essential to select a proper mixing device.

The selection of mixing technology is independent of the method of producing the polymer concretes. The following are the requirements for the mixing devices.

The mixing unit should thoroughly blend all the individual constituents to yield a homogeneous mass. All the fractions of the filler should be adequately wetted by the binder and the formation of lumps or the presence of dry patches in the mix is not permissible. The dispersed fibre reinforcement, when additionally added to the mix, should be distributed uniformly throughout the mass and also wetted well.

The unit should be fast working. The hardening process commences when the constituents of the mix come into direct contact with each other. The time available before hardening of the mass should be utilised for thorough mixing of the constituents to a state of a homogeneous polymer concrete mass, compaction and complete processing into the finished product. Any loss of time in the mixer affects the entire time sequence. It should also be remembered that a chemical reaction occurs in the mixer; the mixer should be cleaned at the appropriate time therefore.

Stratification of the filler is not permissible. This is particularly important when materials of different densities such as quartz sand and hollow glass globules are used in the mixture of fillers. Breakage of the filler granules is not permissible, yet is entirely possible when using light aggregates.

The mixer unit should be amenable to rapid and easy cleaning. The remains of the polymer concrete mass in the mixer and the requirement of detergents and other cleaning (washing) agents should be low while the amount of chemical solvents used for cleaning should be optimal. Simplicity and ease of cleaning also ensure the economy of wages of the operators.

The mixer device should satisfy all the requirements of production capacity. For example, when producing castings weighing 5 tonnes, 50-litre batch mixers should not be used since a hundred mixing cycles one after another will then have to be performed. It should also be borne in mind that the unit should not produce more polymer concrete than can be processed in a given time interval.

The mixer unit should satisfy the requirements of environmental protection. For example, some systems of binders call for the use of explosion-proof equipment. Sometimes, the mixer unit is coupled with other processing machinery. Climatic conditions, particularly when working directly at the construction site, exert a definite influence.

Batch mixers are most suitable for producing test batches and a small number of products. The following types of batch mixers are used:

Buckets (12.3 litre capacity) and agitators—these represent cheap tools which are wholly suitable for initial tests and for producing prototypes of articles;

Paddle mixers, apart from being very slow in operation, are difficult to clean. These devices are not suitable when using brittle fillers, due to the generation of intense shearing forces, but are relatively cheap;

Stirrers with oscillating paddles are also very slow in operation and mixing efficiency is not very high, but shearing forces are low;

Propeller agitators are used primarily in the food and aniline dye industries. The operation of these mixers is extremely rapid but there is danger of break-up of brittle fillers by the rapidly rotating paddles in these mixers. However, the paddles may themselves be damaged by the action of coarse fillers and may become prematurely worn out. Expenses on repair of these mixers are therefore high;

Truck-mounted concrete mixers are used for producing polymer concrete, especially when the volume of material required is large as, for example, when lining finished embankments. Inaccuracies of dosing the reaction constituents and also lack of adequate supervision by the personnel often lead to the hardening of the polymer concrete solution in the mixer. Thus situations have arisen wherein the hardened polymer concrete slurry and the mixer itself could not be utilised;

Integrated mixers function rapidly, can be easily cleaned, and do not use paddles or blades. Thus these mixers preserve the fillers in good condition for a long time.

However, all batch mixers suffer from deficiencies which are particularly distinctly manifest when producing polymer concrete. After each process of mixing, they have invariably to be thoroughly cleaned. Moreover, the remains of the materials and solvents after each cleaning have to be removed, which does not always conform to the principles of environmental protection. Binders are used not only for wetting the fillers, but are also needlessly used for wetting the mixing drum, paddles, and so forth. Therefore, each operation of the mixer requires expensive binders in a larger volume than is required in the composition of a given mix.

Irrespective of the rate of operation of the batch mixers, polymer concrete requires considerable duration for hardening and work cannot be performed on a continuous basis. This is because hardening begins only on completion of the processing of the entire batch of the mix present in the mixer and then the equipment requires cleaning. Therefore, when working with mixers having low hardening periods, batch mixers are unsuitable. In such cases, it is necessary to use continuous mixers which, at present, are already being used for producing cement-based concretes since, in this case also, there is rapid hardening of the mixes as a result of using chemical additives with the cement.

The batch method of mixing calls for periodic dosing of all the individual constituents of polymer concretes. Most often, dosing is manual and hence accuracy and uniformity of dosing depends on the experience of the personnel. Errors in dosing do arise from time to time but they are detected far too late to take any corrective measure. Therefore, in plants operating on a batch system, hardening often occurs either very late or not at all. Often, the reaction commences at the moment when the mix is right in the mixer itself.

Continuous production of polymer concrete products using batch mixers is practically impossible.

Continuously working dosers, mixers, and casting equipment meet the requirements of large-scale production as also for producing special products against special orders. These machines are used not only as mixers, but are also capable of ensuring continuous and accurate dosing of all types of raw materials required for producing polymer concrete slurries so that all the technological operations, i.e., dosing, mixing and casting the products, can be combined into a single technological process. This eliminates time losses and ensures high accuracy and uniform operation.

A vital deficiency of casting machines compared with batch mixers is that they involve very high capital investment which, however, is compensated by several advantages:

—The individual constituents of the mix are dosed mechanically and uniformly; the selected individual constituents of the mix are added continuously with no deviation from the prescribed feeding procedure; human error is eliminated and the machine yields a homogeneous polymer slurry all through the technological process;

—All the dosages of the various constituents of the mix can be regulated, each composition can be easily reproduced and repeated; even the reproduction of a precise colour shade is possible as, for example, to match with polymer marble or polymer onyx after a long interval;

—Within the framework of the technological process, all the operations can be carried out at strictly defined time intervals, the latter being maintained very accurately irrespective of changes in climatic conditions, which exert a definite influence on the course of the reaction;

—The optimum properties of polymer concretes are possible only when the hardening is of high quality; casting machines with a high rate of mixing help to ensure rapid hardening of the mixes resulting in thorough polymerisation of the polymer concrete and in many cases eliminating the need for thermal processing;

—Continuous production considerably enhances the economic efficiency of producing finished products while the use of casting machines helps considerably in reducing the number of workers;

—As long as the casting machine functions in a continuous regime, there is no need to clean it. It is subjected to cleaning only in the case of a prolonged stoppage but even then the cleaning process takes only a few seconds using a minimum quantity of detergents. The use of casting machines thus helps to economise considerable resources and ensures (with minimum wastage) effective environmental protection; and

—The various types of raw materials used in the casting machines are dosed in closed systems; the operating personnel hardly come into contact with the materials and thus phenomena such as gas accumulations, skin irritation, etc. do not arise.

9.2. Providing the Required Capacity

The technological unit should be designed for the given capacity while the casting machine which ultimately determines the overall product yield should ensure this capacity.

In works producing large-sized structures or applying coatings on large areas, there are no problems associated with using highly productive casting machines since large volumes of polymer concrete can be processed quite rapidly. There are no problems whatsoever in selecting small casting machines (for producing small structures and products). Some difficulties can arise when producing a large number of small-sized products and structures using small moulds.

In such cases, the capacity of the machine should, on the one hand, correspond to the overall required number of products and, on the other hand, should not exceed certain limits since polymer concrete fills small moulds more slowly than moulds with large sections and large charging holes.

Industrial plants of moderate size have been designed for processing aggregate fractions of size 3 to 7 mm since the size composition often used in producing polymer concretes covers this range and corresponds to an average wall thickness of about 20 mm of the products. Much larger units with greater capacity can handle fractions up to 16 mm, and some up to 40 mm granules.

When designing the casting machines, the hardness of the filler plays a secondary role since, on the one hand, modern machines are equipped with wear-resistant mixing components and can process solid material (granite, corundum, carborundum, etc.) and, on the other hand, the mixing process is carried out extremely carefully which poses no problem even when processing very brittle, light-weight fillers. The mixing components have high-strength surface coatings to a thickness of a few millimetres and can be restored as they wear out independent of the rest of the equipment. Therefore, delays associated with the restoration of these components are quite few.

When designing the mixing unit, dispersed fibre reinforcement mixed with polymer concrete mix should also be taken into consideration. When the machine is correctly designed, the addition of any reinforcing material to the polymer concrete mix (for example, fibreglass and carbon filament or shaped steel fibre) causes no difficulty in processing them in the polymer concrete mix.

The composition of the filler largely influences the structural design of the machines. This factor is important for designing the dosing unit. Conforming to the varying physical and chemical properties of fillers, various systems are available at present for dosing light- as well as heavy-weight fillers. Fine-grained and free-flowing fillers should be processed differently than hard material, which often tends to cause congestion in the bunkers.

Fillers of mixed fractions are usually used for producing polymer concretes so that the application of the binder can be minimised. Often, the filler mixture is produced away from the casting machine (specially when using four or more fractions). Then, the casting machine can be equipped with only one container for storing the made-up filler material and a unit for dosing it. This ensures the storage of a certain amount of the filler and its addition in the required dose. When using two or three fractions of the filler, the machine should be provided with an adequate number of containers and dosing units so as to carry out preliminary mixing of the filler components before feeding them into the casting machine. A suitable method is selected in each individual case taking into consideration all the local conditions and after identifying all the constituents of the composition.

9.3. Selection of Binder

Different resins used for producing polymer concrete, i.e., polyester, epoxide, vinyl ester, methacrylate, polyurethane, phenol, furan resins, etc., have specific characteristics. They differ from each other in density, viscosity and reactivity, and require different working temperatures for processing and also varying amounts of activators (hardeners).

Depending on the properties of the materials, various types of feeding and dosing systems of liquid constituents can be used. Specially fabricated gear-type pumps are predominantly used for dosing the resins. These prevent high friction coefficients and shearing stresses. The drive of the pumps can operate on direct as well as alternating current; preference is given to alternating current since such drives are cheap and at the same time less sensitive to voltage fluctuations in the feed supply. When using alternating current drives, the control range is usually 1 : 10 but, by using transformers of alternating current frequencies, even wider ranges of control are possible. To ensure high dosing accuracy and to control the number of revolutions, transformers of alternating current frequencies are recommended. These help to mechanise the control of the number of revolutions and, when required, control automatically the number of revolutions with no time lag.

When using, for example, polyurethane and epoxide resins, the hardeners are processed by the same method as used for processing the resins. Only the capacity of the pumps requires to be matched with the required level of material flow.

When producing polymer concrete based on polyester or vinylester resins, the optimum is to have two activators (catalyst as well as promoter) in the form of independent constituents without mixing them with the resins as when preparing the material manually. When storing the constituents separately, it is possible to regulate their consumption even in the course of producing the polymer concrete. This helps to modify at will the proportion of the constituents in the mixture with allowance for local conditions and operational requirements. When processing these resins, extremely small amounts of activators (about 2%) are added to the resin. Dosing should be extremely accurate and is ensured by using diaphragm pumps working at low pressures as different from piston pumps in which very high pressures are developed. In spite of the fact that diaphragm pumps, compared with piston pumps, operate with less intense pulsations, it has not been possible yet to eliminate them completely. Therefore, pumps in the suction and pressure lines are equipped with additional dampers to prevent pulsations and thus ensure smooth flow of liquid fractions into the system. The automatic control of the flow regime of liquid fractions with optical indicators ensures guaranteed, continuous and uniform flow of activators even at minimum volume of the material without giving rise to the phenomenon of pulsations.

The addition of activators using compressed air from suitable high-pressure reservoirs as practised in the plastics industry falls far behind the present level of technological developments. The flow regulation in this case is not sufficiently accurate while the high-pressure reservoir represents a source of great danger.

When processing methyl acrylate resins, the promoter is often already mixed in the resin composition while the catalyst is added in the form of a powder. For reasons given above, preliminary mixing of the accelerator is not recommended in principle, but even such binder systems can be processed in the casting machines. The hardener powder is fed either through a special powder feeding system into the mixer of the casting machine or (in the case of systems using premixed fillers) into the system of filler mixing. In the latter case, there is no need for a separate mixing system.

When processing epoxide and polyester resins, it is preferable to heat the resin in the casting machine to ensure a uniform product of high quality. For this purpose, the casting machine can be equipped with a heating apparatus working on the principle of heat exchangers and not on the direct heating principle. A heat exchanger provides a large heat transfer surface and hence the temperature difference between the heated apparatus and the heat carrier is not high. This eliminates the danger of sudden local overheating. Oil is used as a heat carrier in the heat exchangers. The oil, in turn, is heated in electric heaters. The circulation system ensures a constant circulation of the hot oil and maintains a constant temperature throughout the heat exchanger block.

The heating of the resin significantly reduces its viscosity, which ensures considerable economy of resin consumption, accelerates its thickening and increases the extent of hardening. All of these promote the product quality. Other resins (methacrylate, phenol, or furan), because of their low viscosity, require no heating whatsoever.

The construction material of pumps, pipe lines, valves, etc. should invariably conform to the requirements prescribed for working under conditions of aggressive chemical substances. While for processing resins and their constituents, the use of standard steel or cast iron is wholly suitable, high-alloy steels should be used for components handling activators. When using furan resins hardened by an extremely aggressive acid, the correct selection of the construction material is particularly important. The material used for fabricating packings and hoses should also be acid-resistant and stable to the action of powerful chemical reagents. Materials such as Teflon, Viton, polyethylene, and polyamide are most frequently used for these purposes.

When fabricating and installing electric equipment and pipe lines, the appropriate requirements and specifications should also be satisfied, especially when processing methacrylate resins: in this case, safety of electric equipment and control system used for storing and handling the raw material and in the course of producing must be ensured. These measures are dictated by the extremely low ignition point of methacrylate resins.

In the industry, economy is a decisive factor. In accordance with the well-known rule that expenses should be the minimum required, casting machines have been fabricated for the specific binders used by a given customer. However, at present, in research organisations as well as in industrial plants, casting machines capable of processing various types of binders are required. Such flexibility is desirable for industries in which different systems of resins are processed and also for use in research institutions desiring to enlarge their capabilities. For example, if a consumer desires to work alternatively with polyester and epoxide resins, it is theoretically possible to use a single machine to process polyester resins which, after cleaning, can handle epoxide resins with no special problems. But, epoxide systems cannot be processed in machines intended for polyester resins since a large amount of hardeners is required.

Depending on the required content of hardeners, casting machines should be provided with dosing devices of various capacities. Optimum flexibility can be ensured when the casting machine has systems of feeding and dosing the various constituents. In this case, a switchover from one system of resins to another is possible without expensive cleaning and without risk of undesirable reactions between the different binders.

9.4. Imparting the Required Colour to the Mix

When producing polymer concretes, two types of colouring agents are used: the main pigment (background) added to the mix for its uniform colouration to produce slurries of yellow, white, green, black, or other colours; and the contrast pigment added to the mix for imparting surface effects (for example, imitation marble, onyx, or other natural building stones).

The main pigments can be mixed into the polymer concretes slurry in a powder or paste form. For handling the pigment paste, appropriate dosing devices are available for feeding one or several pastes. These dosing devices are provided with systems of stepless control which makes it possible to impart practically any colour shade. The required amount of the pigment is determined by the colour saturation of the paste and by the pigments themselves which also contain the other constituents of the mixture. A powder dosing device can be used when adding the main pigment in the form of a powder. This device is installed in the casting machine (it is also provided with a system of stepless control) or the powder pigments can be added during the preliminary mixing of the filler, if the process system has an appropriate device for this purpose.

The various possible methods of adding the main pigments to the polymer concrete mix have certain advantages and disadvantages. In practice, pastes are increasingly becoming popular since the colour and its shade can be easily modified without additional cleaning of the system when suitable equipment is available.

The contrast colours, in principle, are processed in liquid or paste form since they are injected at the end of the mixing process into the prepared polymer concrete solution. This injection is in impulses and the duration of each impulse and the interval between the impulses are controlled very accurately for obtaining almost an unlimited number of shades. By using a programmable control system, it is possible to reproduce a given shade at any time (even after a very long period) so as to enable the manufacturer of decorative articles to produce and supply products in the same colour range as claimed. Demands from indentors for replacement on grounds that the products supplied differ from those ordered are almost eliminated.

Pigments in the form of powder or paste can be procured for imparting contrasting shades. In any case, when using them in polymer concrete slurries, the pigments have to be mixed with promoters (not catalysts!). To ensure the pasty condition of the pigment, the powder is mixed with a resin. Powdery pigments have the advantage that they can be stored quite safely for longer periods compared to pigment pastes.

The steadily growing use of polymer concrete has opened up totally new fields of its utilisation, which call for additional processing of additives, apart from the basic raw material going into the polymer concrete mix. The modern casting machines help to process the additives also; for example, de-aerators facilitate the removal of air from the polymer concrete mix during the stage of compaction; special types of paste-forming devices improve bonding between the macromolecules of the filler and the binder; and special devices help to blow air through the concrete to produce foam-polymer concrete solutions.

It should be pointed out that the design concept of the casting machine should take into consideration the solvents which the consumers use for cleaning the system. Not all lining materials are stable to solvents. In some cases, therefore, the hermetic sealing might be damaged and the solvent thus enter the polymer concrete, giving rise to some problems during the hardening of the mix.

9.5. Technology of Producing Polymer Concrete Products

Taking into consideration its long experience with first generation machines, from about 1980 the firm Respecta has been producing second generation machines of three basic models: DB-31, DB-101, and special machines for machine tool manufacture DB-71 with capacities of 30, 100, and 70 kg/min of polymer concrete mix. The control panel of these machines has been simplified and mounted on the upper portion of the machines, which greatly reduces the latter's overall dimensions.

The worm gear system is made of special alloys capable of handling aggregates stones of high hardness right up to corundum and silicon carbide and steel fibres up to 30 mm in length. Moreover, the DB-71 machine is equipped with a special device for cutting glass filaments and feeding them into the worm device. This makes it possible to reinforce the polymer concrete with fibreglass when so required (Fig. 9.1).

Fig. 9.1. Machines for continuous production of polymer concrete mixes: right—machine type DB-31; and left—machine type DB-100.

Compact machines that can be mounted on an automobile chassis are produced for use in road making. These have a capacity of 200 kg/min (Fig. 9.2).

Moreover, the firm Respecta has designed and is producing a series of special mixing machines of capacity 3 to 10 kg/min (Fig. 9.3) and up to 400 kg/min (Fig. 9.4 and 9.5).

Preparation of fillers

The importance of correct selection of fillers and their proper preparation has been emphasised in the preceding chapters. The main criteria are: the fillers and aggregates should be dry and be of correctly chosen particle sizes corresponding to the optimum size composition of the material. Let us recount the possibilities and technological methods used in practice for producing optimum conditions of preparing the fillers (including the size composition curve) for ensuring good technical indices with minimum binder and hence at relatively low cost.

In practice, various methods are available for producing polymer concretes and their advantages or drawbacks can be evaluated from the viewpoint of the available raw material and the demand for the finished product. The following methods represent the two extremes from this viewpoint:

—The supply of the filler mixture in bags according to the order; in this

Fig. 9.2. Casting machines on an automobile chassis designed for applying coatings on the ground.

case, the consumer incurs almost no capital expense on storage and processing of the filler but pays quite a high price for the finished filler; and

—The use of fillers procured from stone quarries or deposits and processed by the consumer's own resources; in this case, the cost of the material itself is relatively low but the capital expenditure on grading and other equipment for processing the raw material into fillers is very high since the stone and other material have to be crushed, ground, dried, sieved and graded.

Between these two extremes, a large number of methods is possible.

Procuring ready-to-use fillers packed in bags. In such cases, capital expenses are minimal but the expenses on raw material go up. This method can be recommended when carrying out tests or for producing prototypes since the volume of the material used for this purpose is usually not much. The danger of stratification according to sizes during transport of the mixture is minimal and hence there is hardly any adverse factor in this regard.

The consumer has only to ensure that the contents of the individual bags are unloaded in the bunker of the casting machine whose feed inlet is located at a height of about 3 m above the ground. When unloading the material into the casting machine from above, a simple chute can be used. When the bags are stored at ground level, simple mechanical transporting devices such as worms or conveyors can be used; the bags are simply dragged manually and emptied into the casting machine.

Fig. 9.3. Casting machine for polymer slurry based on polyester resins: left—capacity 3 kg/min; right—capacity 10 kg/min.

If for some reason the bags of raw material have to be stored for a long period alongside the production equipment, platforms should be erected for their storage. Such platforms do not involve much material consumption and the bags can be stored dry. On opening the bags and loading the filler into the machine, the dust causes no adverse action during the course of manufacturing the products. For mechanical transporting systems, the fillers are loaded into them outside the production buildings.

Procuring different fractions of fillers in bags. This method is more economical than ordering finished filler mixture since material freely available in the market can be gotten in different size fractions. When using casting machines capable of processing several fillers simultaneously, there are practically no additional expenses in producing the mixture of fillers except to take care that a given fraction of the filler is filled in the right bunker of the casting machine.

If the structural design of the machine provides for the use of only one fraction of the filler, the mixture of the filler will have to be prepared from the different fractions before filling it in the bunker of the machine. For these purposes, simple mixing devices like concrete mixers working on the principle of free fall of the mixture are used; the mixture can be stocked on a platform or even a conveyor can be installed.

158

Fig. 9.4.   Casting machine type DB-200 UP/EP used as a detachable system for applying coatings on large surfaces (roads, bridges, etc.).

Procuring the material according to the constituent fractions has certain advantages over the procurement of a finished mixture of the filler. This is not only because of the much lower cost of the different fractions, but also because the consumer has an option in making up a mixture of any size composition conforming to the technological requirements of producing a given product.

Procuring loose filler mixtures in large tonnages in trucks. In some countries, fillers are supplied loose (unpacked) in covered trucks and the material unloaded in large bunkers from which it is later charged into the machine.

Methods are also known of transporting filler mixtures in open dump trucks and unloading them in bulk at the consumer's premises. Since the material, already dried before, has to be preserved dry, such methods of transporting the fillers are possible only over small distances between the supplier and consumer and suitable in regions with relatively stable climatic conditions.

Nevertheless, such methods of transporting fillers are not recommended on a large scale since there is serious danger of the mixture stratifying during transport as also during unloading operations. In any case, such methods of transport do not provide full guarantee of a filler mixture of the required size composition.

Fig. 9.5. Special casting machine in several sections; explosion-proof fabrication.

Procuring individual fractions of fillers in large trucks in unpacked form. This method is extensively used for producing polymer concrete since the consumer (i.e., the producer of polymer concrete) often does not have his own stocks of fillers and the procurement of individual filler fractions helps him to maintain a rational balance between the expenses in procuring the materials and the expenses on storage and processing them.

The individual fractions of a given size composition of the filler are transported separately in trucks (generally in closed hoisting trucks) which helps the material to be maintained in a dry condition and later unloaded into special bunkers of the consumer using pneumatic systems. The design of such large bunkers should meet the following requirements:

—The filler should not be stored in bunkers for long since there is danger of extreme saturation of the material with condensed moisture;

—In regions with extreme climatic conditions, it is most desirable to install filler bunkers inside buildings so as to protect the filler from extreme atmospheric conditions and also to store the material as far as possible at constant temperature. If for local reasons, the bunkers are placed in the open, provision should be made for heating the bunkers using local material. The insulation of individual bunkers is technically wholly feasible but is often uneconomical because of excessive expenses. Experience in producing polymer concretes in the northern European countries has shown that uninsulated cement bunkers placed even in the open air ensure the production of polymer concrete under summer as well as winter conditions;

—The bunker size should be such that it always has a quantity of the filler adequate for production and also for unloading and stocking newly arriving batches of the material; in practice, the bunker volume should be about 1.3–1.5 times the capacity of the trucks supplying the material; thus, if the capacity of the truck is about 15 m^3, the bunker volume should be 20–25 m^3;

—In countries where regular or timely supplies of material are not possible, larger bunkers should be provided so that the consumer can have much larger reserves.

The individual size fractions of the material stocked in the different bunkers is later withdrawn for processing. When the casting machine has many containers for fillers, each bunker should be provided with only one simple device for transporting the appropriate filler into a predesignated container of the casting machine. Dosing and mixing are done later in the casting machine itself.

If the casting machine is designed to use only a single filler, before the casting machine is filled the various fractions of the filler should be premixed according to the prescribed formula in a concrete mixer to the required size composition. Dosing can be done by volume or by weight. Volume dosing is generally cheaper but not as accurate as the weight method, especially if the filler consists of finely ground (powdery) materials, some of which during transport become unevenly distributed. Nevertheless, there are examples (quite a few) of well adjusted dosing by volume.

For transporting the dry filler fractions for producing the composite filler for use in the polymer concrete, belt or worm conveyors are mainly used in the process stages. It should be borne in mind that worm conveyors wear out more rapidly than belt conveyors, but the latter are far more expensive than worm conveyors (because of the need for acquiring and installing special housings for the belt conveyors in view of dust generation and to ensure that the material remains in a constant dry condition). On the whole, however, preference is given to worm conveyors since they are easier to set up; moreover, these devices wear out after a reasonable service life as their usage is limited. Suitable worm devices are also available for vertical transport of material; hoisting buckets and bucket elevators can also be used for the latter purpose.

The individual fractions can be mixed into the final size composition by batch method using mixers of induced type as well as worm devices, i.e., in a continuous regime. There is no doubt that the efficiency of mixing the fractions in an induced type of mixer is more than in the direct-flow worm conveyor but the former involves considerable expense. Both methods of mixing the fillers are widely used in practice and can be recommended keeping in view the production requirement. The use of one or the other method depends on the objective and the available production area.

Fig. 9.6 shows the device of the firm Respecta (type Gazus 249) as a possible design variant for preparing the filler. This unit functions in the following technological sequence.

Four different fractions of the filler are stored in four different bunkers of capacity 30 tonnes each. From each bunker, the fraction enters the weigh-batcher on a belt dosing device. The filler from each bunker comes onto the balance in a sequence and ultimately the required amount of each fraction enters individually the bunkers of the weigh-batcher. This bunker is emptied from the bottom into the bucket of a belt conveyor which, in turn, delivers the required amount of the filler into the batch mixer. After brief mixing in the mixer, the finished homogeneous mass of filler enters the bunker of the casting machine through the exit chute at the bottom of the mixer for later mixing with the binder. The unit for producing the composite filler in a batch mixer is also equipped with a pigment batcher which adds the powder to the mixture of fillers. This helps to ensure uniform colouring of the filler.

This technological process is most optimal for producing large quantities of polymer concrete with a uniform colouration. When it is required to use a different pigment, production has to be stopped for thorough cleaning of the mixer, which poses no serious problem since the mixer contains only dry material but nonetheless does involve some expense. When frequent colour changes of polymer concrete are called for, it is recommended that the pigment be added to the filler (in the form of powder or paste).

The unit for filler preparation type Gazus 467 operates on a different techno-logical flow sheet (Fig. 9.7). This has also four bunkers. Small worm conveyors serve to transport the material from the bunker. Further, dosing (by volume) of individual fractions of the filler and their supply to the mixing worm proceed simultaneously. The finished mixture of filler from the mixer goes directly into the receiving bunker of the casting machine.

Provision is also made to fill in the bunkers when the material is not brought in autoelevators but in open (side opening) trucks or dumpers which do not permit pneumatic transfer of the material into the bunker. In this case, the material is transferred from the dumper into a chute from where the filler, with the help of a feeder, drops onto a vertical conveyor (bucket elevator). A worm distributing system is placed under the discharge chute of the elevator through which each of the filler fractions is delivered into the designated bunkers.

Supply of wet fractions of filler in automobile trucks. Often, individual fractions of the filler can be supplied in a wet state. In this case, the consumer is responsible for drying of the filler. This is best carried out before filling the individual fractions in the designated bunkers (Fig. 9.8) of the unit for filler preparation manufactured by the firm Respecta (type Gazus 544).

Wet fillers are usually transported in open dumpers. The filler is unloaded directly in a chute or (when it is not a dumper) is transferred with the help of a bucket conveyor. A belt feeder takes the filler to a drying drum from where

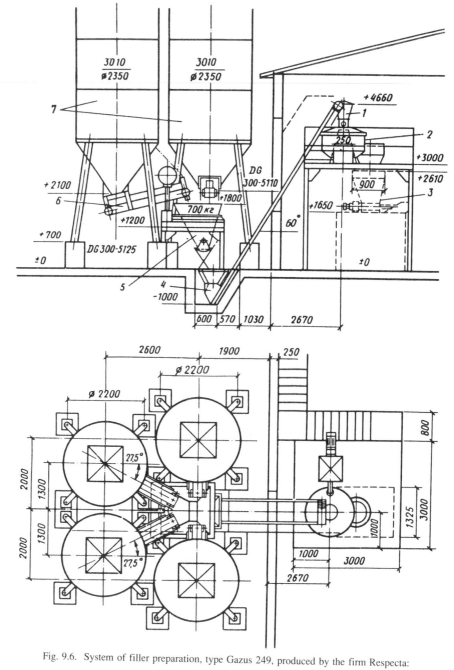

Fig. 9.6. System of filler preparation, type Gazus 249, produced by the firm Respecta:

1—powdery pigment; 2—mixer; 3—casting machine; 4—inclined conveyor; 5—quantitative doser; 6—conveyor-doser; 7—four bunkers.

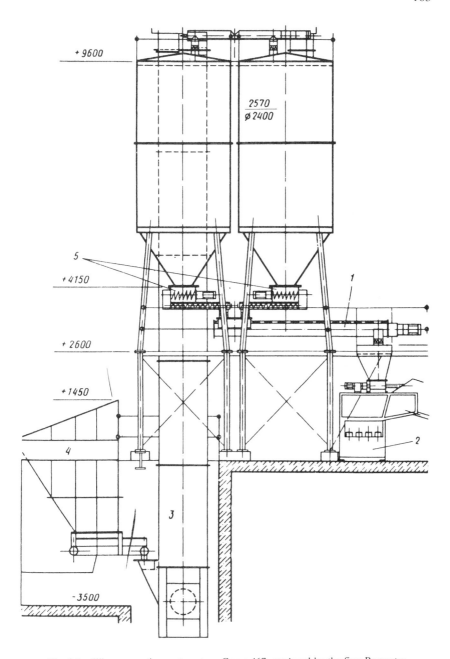

Fig. 9.7. Filler preparation system, type Gazus 467, produced by the firm Respecta:

1—mixing worm; 2—casting machine of the firm Respecta; 3—vertical conveyor; 4—loading funnel; 5—batchers.

164

Fig. 9.8. System for filler preparation, type Gazus 544, manufactured by the firm Respecta:

1—bucket elevator; 2—platform for technical servicing of the unit; 3—belt-bucket elevator; 4—four bunkers each of 25-ton capacity; 5—four microdosers; 6—continuous mixer (flow type); 7—casting machine produced by the firm Respecta; 8—drying drum; 9—feed bunker; 10—platform for servicing batchers and direct-flow

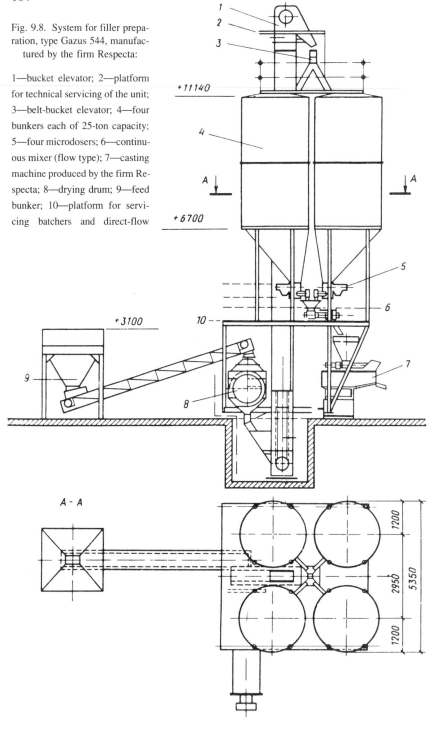

the dried material is discharged onto a vertical conveyor (in this case, a bucket elevator). The worm distributing system is placed under the outlet of the elevator and carries the different fractions of the filler to the designated bunkers.

The individual fractions of the filler are filled through volume batchers and mixed in a continuous mixer to obtain a homogeneous mixture which then goes to the receiving bunker of the casting machine.

Supply of ungraded filler. In many cases, the filler received by the consumer (in dry or wet form) is of indeterminate composition, i.e., the filler may contain large, medium, and fine granules whose ratio is not known. A filler of uneven-size composition may have an adverse effect on the production process of polymer concrete due to uneven quality of the polymer concrete slurries and hence may lead to uneven quality of the product. Therefore, the consumer should adopt appropriate measures for grading the filler at his own expense. For this purpose, a separating elevator (an interative filter) can be used in the system of filler preparation (Fig. 9.9).

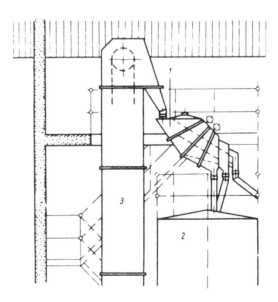

Fig. 9.9. Separating elevator (iterative filter);

1—iterative filter; 2—bunker; 3—vertical transporting system.

The separating elevator is installed between the outlet of the vertical conveyor and the inlet into the bunker. It grades the filler according to fractions and separates these fractions which are very large for further processing.

Ungraded filler material as received may contain only small quantities (or none at all) of fine-grained fraction (mineral flour) required for producing the polymer concrete. Such fine fractions are therefore added to the mixture for preparing the finished filler. The consumer procures mineral flour (in packed condition) or produces such fractions at his own expense by installing grinding mills in the system of filler preparation. These measures pose no serious technical problem but call for additional investment since such mills are quite expensive.

Use of filler material produced by captive stone crushers. In this case, stone breakers operate parallel to the polymer concrete production plant. In other words, stone breaking is an independent economical unit whose output not only goes into the production of polymer concrete but also to meet the other requirements of the firm as well as for selling to other firms. A more detailed description of such stone breaking units is not required here since this is a well-known independent technology.

As already pointed out, any system of producing the filler should be adopted to suit the local conditions. The structural design of the unit should match with the technical conditions laid down and the filler quality required. There are several others besides the structural designs shown in Figs. 9.6 to 9.9. For example, various systems can be combined or some excluded. Also, every manufacturer can design or acquire a unit more fully satisfying his requirements.

Storage and procurement of liquid chemical reagents

It is very important that every consumer provide accurate descriptions and instructions to the manufacturer and supplier of chemical reagents and also and invariably the appropriate standards and instructions of the controlling organs in the corresponding countries. Since the instructions enforced in various countries differ, no detailed descriptions are given here for the storage and procurement of liquid chemical reagents; only the technical conditions essential for ensuring the continuous production of polymer concrete are discussed.

Synthetic resins. Synthetic resins are usually supplied according to the agreement with the supplier in tank trucks of capacity 10–20 tonnes from which the resin is emptied into the containers of the consumer by pneumatic or built-in pumps. The unloading pipe lines are installed in such a way that they stand completely empty at the end of the unloading operation. If the pipe lines enter the container, a siphoning effect can be anticipated.

When pipe lines of long length are used to connect the tank truck with the container or when it is necessary to overcome large height differences, the diameter of the pipe line should be 80–100 mm and sharp angles and bends should be avoided.

To ensure an optimum regime of raw material supply and to prevent interruptions in supply, the collector capacity should be not less than 150% of the volume of supply. To prevent the mixing of the remnants of the preceding batch of raw material with the new batch, an alternative solution is to install two containers, each holding the volume of normal supply.

Reservoirs made of alloy steel or pure aluminium are most suitable for storing synthetic resins. Steel containers are well recommended when the interval between the fillings is not longer than a few months. The containers are usually filled through the opening at the bottom when the gradient is normal. For measuring the liquid level in the container, devices using floats, lever indicators, or air bubble indicators are used. A thermometer is also recommended for measuring the liquid temperature, this being particularly important when the containers are placed in the open air.

To ensure more prolonged storage, material like 'Leguval' should be protected from light and heat. The most optimum storage temperature is 20°C with an upper limit of 25°C. At 15°C, increased viscosity of medium and highly viscous material may cause certain difficulties when emptying the containers and also when processing the material.

To ensure a stable temperature of the resin, the reservoirs should be placed in covered buildings with temperature control. When the containers are placed in the open, effective thermal insulation must be provided (to prevent any excessive cooling of the containers also) so that even when there is a temperature drop, the resin can be pumped through the pipe-line system to the casting machine. Further, the cross-section of the pipe line should take into consideration the viscosity of the resin handled and should be large enough for the travel of the material. The highly viscous epoxide resin, for example, requires large-sized tubes compared to the non-viscous liquid methacrylate resin.

Instead of one large reservoir, several small containers or some storage tanks can be used. These can be connected to the pipe line feeding the casting machine. When using several containers, each may be placed close to the casting machine and connected by a short length of pipe.

Activators

For storing activators, the appropriate instructions and guidelines should be strictly adhered to. Hardeners for epoxide, polyurethane and other resins are essentially regulated by the same requirements and recommendations already given while describing the resins. However, when considering these aspects for other systems of binders, some additional requirements should also be taken into consideration.

Fairly strong acids are often used as hardeners, for example, for furan resin. Therefore, the pipe lines for their storage and transmission should be made of high-quality alloy steels.

When processing polyester resins, the catalysts (liquid peroxides) and accelerators used for hardening should be stored separately. Large volumes of peroxides (over 200 kg) are stored in individual bunkers. The bottom should be made with some inclination in the form of a sump with a settler capacity of about 30 litres. Under certain conditions, the peroxide catalyst can be completely drawn from it. The bottom surface should be smooth and dense and must be amenable to easy cleaning. The closing doors made of steel sheet should open outwards. Slightly above the bottom, air intake lines of 30 cm × 30 cm with louvres should be provided in the side walls of the container. Provision should be made under the lid for additional air intake openings in the required numbers.

The internal dimensions of the storage building should be adequate for total operational safety with various types of containers in which peroxide is handled. The height of the building should be not less than 2.5 m. The capacity of the buildings can be raised by constructing stands with aluminium sections. Aluminium grills can be used as supports; they promote air circulation. Heating the buildings is usually not required in regions with warm climatic conditions; in fact, it is often necessary to provide cooling systems since the peroxide begins to boil and liberate gases at a temperature of roughly +28°C, which adversely affects their activity.

Some storage buildings for activators should be placed in such a way that the containers required in the process, i.e., those used while operating the casting machines, can also be located in them. Then, there will not be a single container with peroxide outside the storage house (with the exception of small amounts of peroxide in the pipe lines connected to the casting machine).

Other chemical reagents. All the other chemical reagents and raw materials required for producing polymer concrete, for example, the separating lubricant, solvent, and pigments, can be stored together. However, they should be classified as hazardous and others, and stored in two different buildings. Normal temperature (not excessively high) should be maintained in the buildings.

Production and use of moulds

Finished polymer products are usually produced by casting, which requires the use of moulds in which the liquid or paste-like polymer concrete slurry is cast and hardened. In designing the moulds for polymer concrete products, the same aspects as in other casting processes should be taken into consideration.

Mould section. Moulds should have openings through which the polymer concrete slurry can be quickly filled. Further, the shape of the mould and the surface properties of the mould material should ensure rapid flow of the solution within the mould. Only by fulfilling these conditions is it possible to ensure an economic production regime of polymer concrete products. Too many narrow inlets in the mould lead to clogging of the mould inlet and adherence of the polymer concrete to the edges of the mould during continuous operation of the casting machines. This causes not only loss of material but also much contamination of the production areas. Very narrow mould sections may disturb the uniform distribution of the polymer concrete mass throughout the mould and result in poor-quality sections in the finished products.

Single sections of moulds prevent temperature differences during the exothermal process of hardening and hence differences in thermal and shrinkage stresses which may lead to the deformation of finished polymer concrete products.

When making moulds, special attention should be paid to complete removal of air from the mould and polymer concrete mass during the process of filling and during compaction of the polymer concrete mass. Depending on the shape of the product and the section of the outlet, it might be useful to provide a few openings in the moulds for air release as in the case of moulds used for metal casting.

Striking (demoulding): For efficient production of polymer concrete and also in view of the rapid reactivity of polymer concrete mass, the process of striking should be automated or (when working manually) the proportion of manual work reduced to a minimum. The use of fasteners in the form of spring tension members, clamps or holders is recommended as their operation is very simple and quick compared with screwed joints. There is no doubt that the basic requirement is to ensure stable and firm adhesion of the constituent parts of the mould.

Striking is also simplified and improved by slanting the sides of the mould slightly inwards. This is particularly important when using monolithic and three-dimensional moulds into which the polymer concrete mass can be pressed in a hot state during hardening. The slope of the mould should be not less than 3°. A higher conicity helps obtain even more effective results.

When the moulds consist of several parts, there is no danger from hot pressing but, in this case, striking should be carried out in stages.

Whenever possible, the mould should be designed in such a way that striking of the inner mould can be carried out first followed by the outer mould. Thus, the outer mould performs an additional function of protecting and strengthening the product which so far (by the time of striking) has not yet totally hardened. However, the inner mould withstanding free shrinkage has already been removed. At the commencement of shrinkage, the outer mould is very easily removed since the finished product comes out spontaneously from the outer mould due to shrinkage.

When producing products without an inner mould, for example, plates, panels, or other homogeneous massive products, no special problems arise during striking. It is best to wait for the moment when the product detaches itself from the mould and can be removed freely.

The casting mass in no case should flow from the mould onto the outer surface. This phenomenon leads to the formation of 'hooks' during striking that have to be chipped away from the finished product or striking becomes impossible.

Compaction and transport. To ensure the maximum homogeneity and density of the finished product, high-quality compaction is necessary. However, the effect of even the most productive compaction units will be restricted if the moulds are not designed for the appropriate compaction processes or cannot support loads during compaction. It is especially important to achieve a situation wherein, during vibration compaction, forces acting on the mould will also act on the polymer concrete mass. This becomes possible only when the moulds can effectively transmit the vibration forces to the polymer concrete mass. The moulds should therefore be placed correctly on the vibrator and fixed if necessary. When using vibration tables for compaction, the mould should have a smooth bottom surface and, when using vibration pedestals, the mould surface should be uniform at places which rest on supports.

Insufficient rigidity of mould walls may lead to shake up of the mould during compaction by vibration and cause leakage of air into the mould and into the polymer concrete mass. Quite often, this problem arises not when using vibrators but when using suspended vibration compacting devices. Undercompacted masses at the corners and edges of the mould also lead to the penetration of air into the polymer concrete mass and to the leakage of the liquid binder. Both these phenomena lead to edge defects in the finished products and give rise to serious problems during striking, necessitating additional work at the finished stage.

Material for preparing moulds. For preparing moulds, extremely diverse materials are used, such as wood, steel, alloy steel, aluminium, glass-reinforced plastics, synthetic rubber based on silicone, polyethylene, Teflon, etc.

For producing special products, prototypes, or a small number of articles, wood is the most optimum material for making the moulds. Wood is a very cheap material compared to metal or plastics but is inferior to them in its durability and hence is less suitable for large-scale production of polymer concrete products. For producing decorative finishing plates and panels, splint-slabs are well recommended for making the moulds (melamine resins are used for coating) since moulds made of these materials are not damaged in the course of use or during striking. The alternative materials, for example corrosion-resistant steel or plastics, are not only more expensive, but also generally very heavy. They should either be relatively thick or have stiffeners to avoid warping.

The surfaces of the finished polymer concrete products produced represent an accurate copy of the mould surface. Products with a lustrous surface are produced in moulds whose surfaces have been polished by an appropriate method. The polishing of wood is practically impossible while polishing of steel is expensive. Decorative articles with lustrous polished surfaces are produced in moulds made of plastic material reinforced with fibreglass which maintains the surface lustre for a long time and, when required, can be additionally polished. Such moulds possess an additional advantage in that they can be copied from any finished product or from a model. Thus moulds of practically any shape, without joints that would leave a mark on the finished product, can be produced.

Aluminium moulds are used for producing a large series of polymer concrete articles, primarily in such cases when the moulds themselves are produced by casting. Considerable expenses are involved only for making the model. From this model any number of moulds can be produced later without much expense. The use of aluminium moulds has yet another advantage, namely, the casting skin is not completely removed from the mould and supports the action of the separating layer (lubricant) and facilitates striking. It is well known that the hardening of polymer concrete is an exothermal process. This fact should invariably be taken into consideration when preparing the moulds. The greater the wall thickness of the product, the greater the proportion of binder in the polymer concrete mass and thus the greater the amount of heat liberated during hardening. The temperature can go up to 50–60°C (which is wholly permissible) and hence the mould should be heat-resistant.

Defects and some problems arise also when the moulds are made by gluing several components together in spite of using glues of appropriate thermal resistance.

For commercial production, the most suitable moulds are those produced in such a way that the polymer concrete products are obtained in a single operation. This involves a certain amount of complexity and expense in mould preparation but these are compensated by eliminating the additional processing of moulds as, for example, for producing products of complicated shapes. However, during the production of prototypes or a small number of products as, for example, when producing large-sized members and structures, it may be technically and economically effective if such structures are produced not in one single operation in the form of monolithic castings, but are divided into constituent members and made in small or less expensive moulds and later assembled into the finished structure. The most optimum technology of joining the constituent members into the finished structure is screwing or gluing. Holes can be made or if necessary provision for screwing in the threads can be made in the body of the mould and cast with no special problem.

Separating layer for mould. Depending on the material used for making the moulds and also the required surface quality of the products, it is necessary to apply a separating lubricant to prevent the sticking of the binder to the mould and thus ensure rapid striking. The separating material is selected after suitable investigations before producing the polymer concrete product. Many companies producing polymer concrete once underestimated the importance of the separating layer and only adverse experience compelled them to recognise in full measure the importance of this material. The presence of cracks and breakage of bits from the finished articles are not merely the adverse consequences arising from the absence of separating layers when producing polymer concrete products, but also result from incorrect selection of this material. An incorrectly selected separating material (or ineffective separating layer) may break the mould right in the first attempt at moulding the product.

The separating layer is intended not only to prevent the sticking of the binder to the mould surface, but also to cover up any possible pores on the surface of the product, cavities in the moulds, slits at the joints, and also to prevent the penetration of the binder or the polymer concrete mass into these cavities. The coarser and more porous the material from which the mould has been made, the greater should be the viscosity of the separating layer and the thicker the separating material layer applied on the mould. Thus, moulds made of wood are treated with wax before filling with the polymer concrete mass; wax also functions as a separating layer. But the surface of the casting cannot be expected to be lustrous in this case.

A lustrous surface in moulds calls for the use of very thin separating layers. For this purpose, polishing wax or pastes are used (after applying, they are almost completely rubbed in); liquid separating materials are also capable of imparting some lustre to the mould surface. There are also some so-called 'internal separating materials', i.e., material is added to the binder itself which prevents its adherence to the mould surfaces and eliminates having to process the mould surface directly with a separating material. However, such materials are not well recommended in practice. A definite separating effect is possible only when an extremely thin layer of the internal separating material (gel coat, a gel-forming layer) is deposited. Whether such a separating layer is adequate for ensuring a high-quality article and whether it has a positive effect on the mould, it is difficult to predict. In any case, there is real danger of insufficiently effective separation and hence the breakage of finished articles during striking cannot be eliminated.

Other criteria for selecting the separating material are thermal stability and rapid drying, for example, by the action of air. The entry of the solvent present in almost all the separating layers onto the upper layer of the polymer concrete or deep into the mass should be prevented, as it may cause difficulty during hardening and also visual defects on the surface of the polymer concrete. Moreover, the separating material should be firm enough not to be dislodged from the mould surface under the action of the polymer concrete mass; otherwise it will be practically ineffective during striking. This danger is particularly imminent when coarse fillers are present.

The desire to develop a universal separating material that could be used for all mould materials and for all binder systems, functioning effectively and uninterruptedly, is quite understandable, but such a material is hardly possible. A separating material which is closest to all these objectives is polyvinyl alcohol, a watery liquid forming a film on the mould surface which acts as a genuine separating layer. The main advantages of polyvinyl alcohol are: it can be used on almost all the materials from which moulds are produced; even while applying the material on the mould surface, the quality of the separating layer can be visually judged since, for example, on surfaces containing fats it is generally impossible to produce a good and compact separating layer and the polyvinyl alcohol will actually flow off the mould surface; the application of a uniform separating layer makes for ease in striking; polyvinyl alcohol is available in various shades and hence a grade of alcohol matching with the colour of the mould surface can be selected and, finally, the phenomenon of binders sticking to the mould can be avoided.

The main difficulty in handling polyvinyl alcohol arises when it is applied in an excessively thick layer that requires a very long period of drying. Polyvinyl alcohol contains a significant amount of water and its hardening takes considerable time (only a hard layer can withstand the action of a polymer concrete mass cast in the mould). The thicker the layer of polyvinyl alcohol, the longer the period required for drying. It would be incorrect to assume that because the surface layer of the polyvinyl alcohol hardens rapidly, the underlying layers too have dried; in fact, the deeper layers take a relatively longer time for drying. The hardness of the layer can be felt with the fingers, giving an idea of the state of the surface layer. Therefore, it is very important to apply polyvinyl alcohol in an extremely thin layer (to the extent possible) so that it dries well under the action of air and hardens.

Moulds of polyethylene, Teflon, and similar materials can be used entirely satisfactorily with no separating layer. The use of such materials is limited, however, because of their high cost.

Compaction of polymer concrete

When compacting a polymer concrete mass, the very same principles as in other similar fields, for example, in cement concrete production, are applicable. Polymer concrete can be compacted by compression, vibration or centrifugal force (as done when producing products having a circular section, for example, pipes), or also by centrifuging or compression by rollers. It is important that the viscosity of the polymer concrete mass matches totally with the selected compacting technology. Only then are positive results possible.

However, in practice, preference is given to vibration compaction since this is a universal technology well recommended for materials ordinarily used for producing polymer concrete. The criteria which should invariably be taken into consideration during vibration compaction of polymer concrete are given below.

In the process of compaction, the following principle is *not* applicable: 'The longer the vibrations, the better the compaction of the mass and hence the better the quality of the finished product.' It is essential to recognise and keep in mind that vibrations over a prolonged period are fraught with adverse consequences for the end product. Very heavy and coarse fractions of the filler descend more rapidly than the much finer and lighter fractions and hence there is danger of stratification throughout the section and, since the much smaller fractions of the filler have a greater specific surface, the upper layer of the mass is bound with a greater amount of the resin than the lower portion consisting of coarse fractions of the filler. The uneven distribution of the binder throughout the mass makes for uneven shrinkage, which might lead to warping and bending of the finished product. This danger is particularly high when producing large-sized products, for example, plates (slabs) or panels.

Compaction may also have an adverse consequence if vibrations are prolonged after gel formation of the polymer concrete mass has begun, i.e., at the commencement of the hardening reaction. If in this case, vibration compaction is continued, there is danger of decompaction of the mass, leading to a heterogeneous polymer concrete mass.

When producing polymer concrete products used for decorative finishes and capable of producing colour effects, vibration compaction may exert a definite influence on the outer appearance of these articles. Colour effects are imparted mainly in the course of filling the moulds with the polymer concrete mass and are largely dependent on mould movements. It is difficult to judge to what extent such movements have a positive or adverse effect and much depends solely on personal judgments born of experience.

In polymer concrete production, vibrating motors with pneumatic or electric drives are used. The advantage of vibrating systems with pneumatic drives is the fairly simple control of the frequency of vibrations by reducing or increasing the air supply. The drawbacks of these systems are the extremely insignificant oscillating forces (amplitudes), high level of noise during operation and the large consumption of air. Therefore, a large number of consumers are more favourably inclined to use electric drives, usually without a frequency control system since the latter is expensive. The consumer generally produces compaction at normal (about 3000 rpm) or at maximum frequencies (about 6000 rpm). Further, several cases are known in which the vibration tables, used for compaction, were operated alternately on normal or high frequencies. Which level of frequency is most optimal is a question to be examined for each individual case. As a rough guide, the following indices are useful. When producing decorative members using a polymer concrete mass with fine fillers, in almost flat moulds of low height usually high frequencies are used. When producing engineering products in metal moulds designed for a limited weight and size, work is usually carried out at normal frequencies.

The vibration force should be adapted in each case to the mass of the casting and the mould. When using vibration systems with an electric drive, this poses no complexity and requires only the control of the eccentric discs.

The most popular method of transferring the compacting force to the polymer concrete mass is to place the mould in the process of filling it with the polymer concrete on the vibration table in such a way that the force applied acts downward. Vibration tables are universal; in other words, using vibration tables, the mass can be compacted in almost any mould. Instead of vibration tables, vibration pedestals or vibration frames could also be used but the latter should be adjusted more accurately to the dimensions of the finished moulds. In a continuous production regime, two moulds are generally placed on the vibration table at the same time, one each at the beginning of filling and at its completion (i.e., after filling the second mould), and thus two moulds can be compacted simultaneously.

Instead of vibration tables, suspended vibration systems may also be used. These are directly connected to the mould with screws or clamps, for example. This method is well recommended only when producing large-sized products wherein filling the mould is protracted, when working with one or several moulds, and also when working with vibration systems with a pneumatic drive that does not involve much expense, and for producing large numbers of products. The main working conditions required for the latter is the availability of rigid moulds which do not bend inward.

There is also a third method of compacting the polymer concrete mass in moulds, i.e., by using manually operated internal vibrators which are inserted directly in the polymer concrete mass and are withdrawn from the mass in good time just before the commencement of hardening. This method is suitable mainly for those consumers who use manual vibrators for other purposes, i.e., these are already available. Implementation of this method requires an additional power source since these vibrators are controlled only manually. The use of vibrators is also limited to the section of the mould inlet. Manual vibrators are also appropriate when used as ancillary vibrators whenever the main process of compaction is done on a frame, pedestal or table. Later, sections of the mould in which the polymer concrete mass requires special compaction, can be compacted using a manual vibrator.

The compaction of a polymer concrete mass upward, i.e., with the help of vibrators placed under the mould, is not recommended. The work of the vibrators in this case is less effective since they cannot produce sufficiently intense compaction. However, such methods can be used as ancillary operations, for example, to obtain a smooth and even surface on the backside of the casting.

Some vital errors that creep in during the compaction of a polymer concrete mass are given below:

—Vibration tables or frames are not sufficiently rigid and hence the vibration forces are non-directional; with insufficiently rigid structures, the vibration energy does not enter the polymer concrete but is converted into kinetic energy of the supporting structure of the vibration table or frame; after prolonged working, this causes fatigue of the material constituting the support;

—Intense noise is produced if the moulds are not properly secured on the table (especially when the moulds or the table are made of metal); with the help of a simple mechanism (pneumatic or hydraulic) controlled by a switch, the mould can be fixed to the table or to any other support surface; thus not only the noise level is reduced but the application of force to the polymer concrete mass can be directed as required and compaction improved;

—Several components subjected to vibration simultaneously also increase noise and wear; this is relevant when guide rolls on which the moulds are transported are also involved in vibration; this can easily be corrected by using vibration pedestals which in the course of vibration are shifted between the rolls with the help of a pneumatic or hydraulic mechanism; the roller conveyor can be placed below the vibrating pedestal; and

—When using several vibrating motors simultaneously on a single mould or on frames of a single vibrating system, special attention must be paid to ensure that the vibrators rotate in the same direction; the different vibrators should supplement each other as otherwise simple mixing of the polymer concrete mass will result without compaction.

9.6. Some Recommendations for Production of Polymer Concrete Products

After filling with the polymer concrete mix and compaction of the mass, the moulds go for further processing of the polymer concrete, viz., hardening of the polymer concrete (in this stage, the moulds can be fully transported); striking; cleaning and applying the separating layer and assembling the mould; drying the separating layer in air (this can also be done while transporting the mould); application of a thin coating (when producing decorative products) on the mould surface; hardening of the thin film coat (can be carried out during transport) and fresh filling of the mould in the casting machine.

When designing a plant for producing polymer concrete products, all the local conditions should be taken into consideration: magnitude of the earmarked area, size of the shops, supply of raw material, equipment, etc. These conditions naturally differ at different places.

When designing a plant for producing polymer concrete products, its dimensions are determined in relation to the required production capacity, sizes of the moulds used, and also the time required for carrying out the various stages of the technological process. Significant differences also arise, for example, regarding the separating layer used, i.e., standard or 'gel coat' (thin film layer). This factor alone (which, in fact, involves significant time considering the time required for hardening of the thin film layer) can either doubly lengthen the production process as a whole or reduce it by half. The longer the duration of each technological cycle (i.e., the longer the interval before the same mould goes for the next filling), the smaller the number of moulds and the area required.

The same phenomenon also applies to the life of the polymer concrete mass. The lower its hardening rate, the longer it resides in the mould, and therefore the longer the duration of the production cycle.

It is entirely possible to reduce significantly the hardening duration of the polymer concrete mass or the thin film layer by organising thermal processing, but this involves additional capital investment and power consumption.

The usefulness and hence the economy of organising thermal processing depends in a given case on the climatic conditions. As already pointed out, a stable temperature of around + 20°C is regarded as optimal; since common systems of binders are quite reactive at this temperature there is no need for additional heating.

From the time required for one complete cycle of casting a product in the mould and the next filling of the mould with the polymer concrete mass, the optimum number of moulds can be determined to ensure continuous production and thus optimum utilisation of the production capacity.

The devices intended for transporting empty or filled moulds should be designed taking into consideration the moving members. Here, the adage that expenditure should not exceed the limit of usefulness is justified. When a manufacturer lacks adequate experience, he should be extremely cautious towards the adoption of automated flow lines. Each technological unit comprising individual components requires some time to become fully operational and anyone working with a new device requires a 'run in' (gestation) period. These two problems should be resolved successively and not simultaneously.

Roller conveyors are generally used for moving moulds, and are quite common for almost any type of mould. Other systems of transporting moulds are also available, such as elevators, rotating tables, etc. Moulds can also be secured at one place and the various operations carried out in fixed moulds. In this case, to move the polymer concrete mass from the casting machine to the mould, belt or bucket conveyors can be used.

Castings without internal supporting elements are amenable to easy striking with good compaction and shrinkage as can be observed by examining the edges of the mould. The fixing rods should as far as possible be removed before total shrinkage of the material since pressing and shrinkage might cause a mechanical joint between the mould and the product. In such cases, the actual material used for producing the mould, its dimensions, conicity, etc. should be taken into consideration. When and how best to carry out striking should be determined beforehand based on the design of the mould and the raw material parameters, but this can be established more accurately only in the actual plant practice.

In addition to moulds, raw material and finished products must also be transported during production work.

There are no serious problems whatsoever in transporting the liquid raw materials. They are transported from their storage containers to the production site through pipe lines and hoses right up to the casting machine under their own gravity. The pipe lines and hoses should be laid in the works area in such a way as to ensure free flow of liquids while protecting the system from damage. In the course of designing, the possibility of dust formation should also be taken into consideration as dust not only adversely affects the workers, but can also lead to contamination of the finished surfaces of the products. Therefore, the preparation of the filler and the production process of the finished product should be carried out in entirely different buildings.

When transporting finished products, it should be remembered that they have not completely hardened at the time of striking (only to 50% of ultimate density) and hence, in the period of hardening, they should be protected from high mechanical loads, especially at vulnerable points or sections. Safe handling of the polymer concrete products is possible only after they attain 90% of the final hardness. Depending on the raw material and the reaction period, it requires about 3 h after casting is completed for the polymer concrete heated by the heat liberated during the reaction to cool to room temperature.

Although the material has not fully hardened by this time, it can be safely transported and, when required, can be subjected to further processing. It is quite possible (but not obligatory) to perform such additional operations as cutting, drilling and polishing, grinding of polymer concrete castings, etc.

Since the hardening of the polymer concrete occurs in the individual moulds, it is possible to produce moulds in such a way that the finished articles are obtained from them without further processing. Different moulds may have, for example, different types of surfaces (for producing plates or panels with lustrous or mat surfaces) or different sizes (so that the product may fully correspond to the finished dimensions). The components and members additionally introduced into the polymer concrete can be inserted in the mould even before casting so that their bonding at a later stage can be eliminated.

Recommendations on the production of polymer concrete plates and panels

Plates and panels of polymer concrete (for example, polymer marble or polymer onyx) are used for lining facades, making window sills, benches, table-tops, wall facings, staircases, and floor tiles. The design, colour and dimensions of the products are determined in accordance with their designated use.

For works producing plates and panels, it is important to decide first of all whether to produce panels of standard sizes and cut them later to the dimensions required by the buyer or to produce the finished product conforming to the actual dimensions prescribed by the buyer. The solution to this problem depends greatly on the prevailing type of marketing facilities: when marketing through special organisations, the functioning should be different than when supplying directly to the buyer.

A producer selling his output to a marketing organisation generally produces and supplies panels and plates of standard dimensions and the marketing organisations then cut them to the sizes required by the buyers. If, however, the manufacturer supplies the products directly to the buyer, he himself has to meet all the requirements of the buyer. In practice, the producer has to organise the work conforming to both these methods. Thus, when discharging a large order for panels of given dimensions designed as facing for a building, it is desirable to prepare (or buy) special moulds to meet this order. When the order is large enough, the cost of these new moulds is steadily recouped while additional expenses on otherwise cutting the standard articles into finished measures are eliminated. In this case, the special dimensions become a standard for all practical purposes.

From the technological viewpoint, the design of technological lines for producing products of one given size is relatively simple. Such a production line can be mechanised quite readily and with a minimum of capital investment. When producing products of standard sizes, it is useful to produce panels up to 5 m in length. However, works producing mainly facing materials for interiors of buildings generally limit the maximum length to panels of 2.5 m since this size corresponds to the average height of buildings. In any case, the production of panels to different widths must be organised, viz., from 10 to 60 cm at 5-cm intervals. This helps in selecting the panel width as required by the customer and the panels then need only to be cut lengthwise. The manufacturer should avoid producing plates or panels of very large size since this is more complicated and difficult than producing small ones. On the other hand, large products are less prone to deformations (due to their very high intrinsic weight) than smaller and lighter panels.

As already pointed out, a wide range of materials are available for producing moulds. Splint-slab boards (wood laminates) with a coating based on melamine resins are particularly popular because of cost considerations. This material can be obtained in different finishes: mat-lustrous, lustrous, sections, etc. However, the need for using glued or screwed edges makes possible the production of only rectangular moulds, which is regarded as a drawback, but such shapes are quite common in many fields. When required, the production of special products, for example, with rounded edges, is possible by using plastics reinforced with fibreglass as material for preparing the moulds.

Decorative panels and panels for facing facades as well as interiors of buildings are generally produced after a thin surface film about 0.4 mm in thickness has been deposited on the mould surface subsequent to the application and drying of the separating layer. The thin film layer is applied by spraying or brushing. As in the case of painting, the application of the layer by spraying ensures a more uniform layer offering a greater aesthetic external appearance compared to brushing. However, this involves a significant increase in capital investment, not only in acquiring the painting booths but also in buying the required exhaust equipment for clearing the dispersion (mist) formed during spraying of the material. As soon as the thin film coating applied on the mould has hardened sufficiently, the polymer concrete mass can be cast in it; the filler in the polymer concrete cannot damage the mould as this thin layer performs a protective function.

In the process of compacting the polymer concrete slurry, care must be taken to safeguard the proper duration of vibrations as mentioned earlier, with excessive vibrations the dangers of stratification of the fractions and resultant warping of the panels become particularly acute.

The danger of possible deformations can be greatly reduced by ensuring uniform heating and cooling of the mass during its hardening. A very simple method is to cover the mould filled with the polymer concrete solution after the commencement of gel formation with a material possessing low thermal conductivity. This will prevent the exothermal heat liberated in the course of the reaction of the binder, which is held within the mass by the mould, from escaping through the upper surface of the panel. The panel will thereby be uniformly heated and uniformly cooled, and even retain a smooth surface.

Striking of the panels is very easily carried out after shrinkage of the material when the latter is quite strong. The panels may be recovered from the moulds either manually or by using a hoisting device (vacuum striking) or by overturning the mould and lifting it from the panel.

After striking, it is important to ensure that the panels are kept on an even level right up to their complete hardening to preclude deformation under their own weight. After final hardening of the products, their additional processing (when so required) can be carried out and the panels then stored. When producing panels with a lustrous surface, after hardening, an additional polishing operation should be carried out (in spite of having applied a lustrous thin layer on the mould surface). A simple polishing machine does not involve much capital investment. The expense is more than compensated by the economy of the casting moulds. The moulds can then be used more frequently even if the surface lustre is not quite adequate or minor defects are present on the surface layer.

For transporting the moulds and finished products, a roller conveyor is most advantageous. Conveyors can be set up at different levels one above the other so as to economise the production area and take the best advantage of the shop height.

When producing panels with an area of 1 m^2 and weighing about 30 kg, the following is the usual production schedule:

Filling the moulds and compaction	1 min
Hardening in the mould	20 min
Striking	1 min
Cleaning the mould and applying the separating layer	2 min
Air-drying the separating layer	5 min
Applying the thin film layer	1 min
Hardening of the thin film layer	15 min
Total time required	45 min

The above schedule suggests that a given mould is ready for re-use every 45 min and hence continuous production of polymer concrete products is possible with 45 moulds.

The unit shown in Fig. 9.9 illustrates the statement that, when arrangements are available for heating the moulds, work can be significantly more rapid. The technological unit shown here has been placed in a shop with a total area of 150 m^2 and uses 24 moulds placed on a rotating table. This enables continuous moulding of panel boards of 0.25 m^2 in area weighing about 5 kg each. The casting machine in this case functions in a production regime of 10 kg per min, which makes it possible to cast two panels every minute. The output per shift of this small unit is 240 m^2 of finished boards.

In this case, the various operations are performed to the following time schedule:

Filling the mould	0.5 min
Hardening in the mould	3.0 min
Striking	0.5 min
Cleaning the mould and applying the separating layer	0.5 min
Air-drying the separating layer	1.5 min
Applying the thin film layer	1.0 min
Hardening of the thin film layer	5.0 min
Total time per cycle	12.0 min

Production of sanitary-engineering products

Sanitary-engineering goods are usually produced in plastic (reinforced with fibre glass) moulds so as to attain a lustrous surface. Sanitary-engineering products come in complex shapes and hence moulds consisting of two or more parts (inner and outer moulds) are used in casting them. These fully meet with the requirements prescribed for the finished products and there is therefore practically no need for their subsequent processing. For effectively transferring the lustre from the mould surface to the product surface, high-quality separating materials, i.e., pastes and polyvinyl acetates, are used. The 'gel coat' layer (thin film separating layer) is usually applied by spraying in order to cover even those sections of the mould which are not accessible to a brush. One of the advantages of spraying the thin film layer using a special table which can be rotated in different directions, is that it ensures effective exposure of all the components and sections of the mould to the spray. Professional competence is essential to obtain a fairly uniform thin film layer while preventing the collection of large amounts of the resin in the corners and edges of the mould.

The thin film layer is applied most frequently on those sections of the mould (as a rule, the inner moulds are given such a coating) which later form the external surface of the finished product. The application of the thin film layer on the concealed surfaces may be regarded as a needless luxury. For forming the separating layer on the concealed surfaces, a cheaper material, such as simple wax, may be used. It can be applied rapidly and without technical problems since wax does not call for additional polishing. In the course of hardening, this wax (not heat resistant) melts under the influence of the exothermal heat and therefore the corresponding mould component can be withdrawn quickly soon after the hardening process commences.

An unduly long time should not be allowed for the hardening of the mass in the inner mould since, with time, the danger of the polymer concrete slurry pressing against the inner mould intensifies. Striking the inner mould is best carried out using frames for withdrawing the finished products from moulds. Such frames should be available at every works and match the size and shape of the castings and moulds. The striking frame should be so made that the inner moulds with the finished product in it can be held on the frame while the finished product drops through.

The height of the frame should be established in such a way that the finished product does not fall from an excessive height. Damage to the finished product can largely be prevented by using a soft bedding of expanded polystyrene or foam rubber. For easier recovery of the finished product from the mould, the latter can be gently 'dropped' on the frame since the finished product is readily removed from the mould surface by reverse striking. Further, attention should be paid to ensure that the finished article is not distorted in the mould. The withdrawal of the product from the mould is also greatly facilitated by using special hardening systems which impart some initial strength to the mass even before its shrinakge. One should not overlook the possibility of the sanitary-engineering product becoming deformed under its own weight when the hardening process is prolonged. Therefore, the finished product should be placed on a supporting frame right until it attains adequate strength.

In spite of the high cost of additional polishing operations of the finished product, it is recommended for sanitary-engineering goods at the end of casting. The shine on the surface has no relevance to the quality of the product but it can produce a definite impression on the prospective client.

Production of engineering goods

Products made of polymer concrete are called engineering goods only because of their technical properties. Indices such as colour, aesthetic external appearance, etc. play no special role. Such goods are usually produced without applying the thin film layer in steel moulds made of sheet steel by stamping or welding. The life of such moulds is quite long keeping in view their greater usage for producing a large number of products. Usual pastes (wax) are used as separating layers. These are applied in a thin layer since there are no special requirements whatsoever for the quality and external appearance of the product surface.

In world practice, as already pointed out, troughs made of polymer concrete are most often encountered. The average period of filling the mould for a trough of average length (about 1 m) is one minute while compaction is effected simultaneous with the filling of the mould and the product is withdrawn from the mould after four minutes. The separating layer is applied by spraying in a matter of a few seconds, and, after a short duration of drying in the air, the mould is ready for filling with a new batch of the polymer concrete slurry. Thus, the total duration of one complete technological cycle including transport through the various operations of the process is about eight minutes.

Large-sized polymer concrete products are also produced essentially in the same manner. But the question of the hardening duration and the recovery of the finished product from the mould call for special consideration.

The production of polymer concrete and the products made from it depends on the duration of hardening which, in turn, depends on the time of gel formation. On the one hand, the manufacturer attempts to reduce the hardening duration of the polymer concrete, and on the other, a certain amount of time is essential for filling and compaction of the polymer concrete mass (sometimes sufficiently long). A good example in this context is the oil separator. With a product weighing up to 1500 kg, filling of the mould with the polymer concrete mass (at casting machine output 30 kg/min) takes 45 min, i.e., for the material to commence hardening after being filled in the mould and completion of the compacting process, the mass should remain in the gel state for about 50 min. When using continuous casting machines, the hardening duration of polymer concrete can be sharply reduced by steadily increasing the reactivity of the polymer concrete slurry in the course of filling the mould by increasing the amount of activator. In practice, when continuing to fill the mould for about 45 min, the duration of holding the mixture in the mould and recovering the finished product from it is about 1.5 h considering that the process of recovering the finished product from the mould takes some time in view of its rather large dimensions.

Moulds intended for producing engineering goods should be designed and fabricated in such a way that striking commences with the inner mould. The outer mould thus plays the role of a support for the insufficiently hardened body.

Production of pipes

The technology of pipe production does not involve moving the moulds in the technological cycle but, on the contrary, moving the finished but not yet hardened polymer concrete to stationary moulds for further processing.

Polymer concrete pipes are produced by several methods: centrifugal, vibration or radial formation, and also by rolling in the form of combination (sandwich-shaped) tubes together with plastic materials reinforced with fibreglass.

The pipes are produced by centrifuging the polymer concrete slurry of plastic consistency around a horizontal axis. In this method, the following structure of polymer concrete in the pipe walls is typical: much of the synthetic resin with fine fractions of the filler remains on the inner side of the wall while the much larger fractions concentrate on its outer side. When producing pipes using this technology, only centrifugal force acts on the polymer concrete mass and compacts it. Mechanical loads on the mould are not high since centrifugal forces usually act on the mould surface perpendicularly. The rate of mould rotation is about 18–20 m/sec. The duration of centrifuging the mass after its casting in the mould is about 3 min. After this, a thin lustrous film of the resin is seen on the inner surface of the pipe. The polymer concrete mix should be formulated in such a way that polymerisation commences at the end of this duration and advances in the next 6–7 min to such an extent that the process of mould rotation can cease. An extremely important condition of producing polymer concrete pipes by the centrifugal method is control of the process of vibrations caused by the presence in the mix of a relatively high proportion of synthetic resin, which is essential for the centrifugal process.

To compensate for the tensile stresses arising on the inner surfaces, pipes of large diameter should be prestressed. After some extent of prestressing, it is possible to cope with residual shrinkage and creep. Since the material is of high quality, prestressing in a longitudinal direction should be avoided.

Compaction around the vertical axis is carried out by vibrating machines working on a shank principle and also by radial presses working on the principle of a rolling head. When working with a shank-type vibrating machine, the supporting centre of the shank is placed in the lower portion of the mould at the commencement of the pipe production process. On continuous feeding of the polymer concrete mix into the mould, the vibrating shank is slowly raised upward. Toward the end of the process, the shank can be dropped downwards rather quickly while the mould is raised upwards. The pipe can be removed from the bearing sleeve in a vertical position and hardened.

When producing pipes in a radial press, at the commencement of the production process the rolling head is placed in the lower portion of the mould and is raised upwards while rapidly filling the polymer concrete mix. The press portion with the smoothening plunger is placed on the same axis as the head. Steady compaction along the entire pipe is produced using an automatic compaction regulator which controls the rate of movement of the press portion. Toward completion of the raising of the rolling head, the polymer concrete pipe can be removed quickly together with the mould from the press table and striking carried out. An interesting point in this technology is that the pipe possesses adequate stability even without the bearing sleeve.

The heat of friction generated during compaction on the inner wall of the pipe stimulates the commencement of polymerisation.

An important advantage of both the technological methods of vertical compaction of the polymer concrete mass is the possibility of reducing by half the content of the binder in the mass (with the content of the coarse fraction of the filler remaining the same) compared with pipes made by the horizontal centrifugal process.

During production in shank-type vibrating machines, a significant wear of the mould and the shank caused by the vertical displacement of the coarse fraction of the filler is noticed; wear in radial processes with a rolling head under conditions of forces acting almost perpendicular relative to the mould surface can be regarded as low.

New developments in the field of pipe production have made it possible to manufacture composite (sandwich) pipes, referred to above. Such pipes are produced by rolling technology, i.e., by the method used in the production of polymer concrete from commercial plastics. Further, fibreglass impregnated with a resin or a mixture of resins is applied on the shank while simultaneously adding the polymer concrete mass. This method has been described elsewhere in sufficient detail.

Production of pedestals for machinery and lathes

In mechanical engineering practice, the tolerance requirements are more exacting than in civil engineering and the building materials industry. These requirements determine not only the selection of the raw material (and the composition itself), but also exert a definite influence on the selection of the technological processes.

When preparing the polymer concrete mass, it is essential to conform precisely to the given composition and the homogeneity of the slurry produced on mixing. Further, the homogeneity of the composition should be maintained all through the technological process of producing the articles. This aspect may be regarded as decisive when using casting machines.

The manufacturer should pay special attention to the production of the moulds. These should be made with precise tolerances to ensure their stability so that, during filling and compaction of the polymer concrete mass, all the mould components are cast in accordance with the given dimensions and remain in that state. For this purpose, even before the commencement of filling, the moulds should be levelled on the vibration table and fixed. The force of vibration, frequency, and direction should be controlled in the optimum regime. This, in the ultimate analysis, is possible only when using appropriate materials and compositions. Even then, the polymer concrete mass is fed into the mould and not vice versa except when working with a single mould. The casting machine can then be set up directly above the mould and the concrete mass fed directly into the mould. The polymer concrete is transported using bucket or belt conveyors.

Machine or mill components made of polymer concrete represent load-bearing structures and often tool supports or devices for holding articles, drives, and transport attachments are fixed on them. Sometimes, even bins of distribution systems are fixed on them. Therefore, when producing these structures, provision should be made for guides and for fixing various devices. For fixing the auxiliary units, for example, hydraulic devices and instruments, threaded inserts are provided. For conveyor attachments, to impart strength adequate for taking the loads, concealed metallic bodies or conveyor tubes, for example, of polyvinyl chloride, are provided. Detachable bars are made (depending on the type of concrete) of polystyrene or polyurethane for reducing the weight of the overall mass; their position is determined in such a way that conveyor lines can pass through them without disturbing the rods when casting the concrete. To increase the weight of the mass, the rods may be made, for example, of cement concrete. In rods made of foam plastics, gas pressure might arise and cause stresses in the pedestal. In such cases, provision should be made for openings for air escapage. When fixing headstocks such that the stresses generated are minimal, provision should be made for compensating surfaces corresponding to the resting surfaces of the components that require to be fixed to the polymer concrete product.

Threaded inserts, panels, or planks are made beforehand and fixed to the upper surface of the model. To prevent the polymer concrete solution from falling between the wall of the mould and the inserts, adequate compaction must be ensured. For good adhesion of the metallic components to the polymer concrete, the components should be mechanically cleaned and adhesion activators applied on their surfaces. Additional anchors may also be considered for panels and boards. It is possible to install guide channels also. They can be integrated by casting with the polymer concrete mass or precast panels can be used. The semi-finished material, before the hardening of the polymer concrete mass should be processed to obtain a uniform and smooth surface and provide tight adherence to the mould walls. Additional strips may be used for compaction between the mould and the guide channel. Later, the polymer concrete is cast in the mould and joined to the metal surface using strong clamps. A drawback of this system is the danger of stress generation in the polymer concrete and in the guide channels caused by the temperature of polymerisation and the varying coefficients of thermal (linear) expansion after the polymer concrete hardens. This can be avoided by gluing the guides or screwing them from above on separate boards.

Technology of applying coatings

The main condition for ensuring the durability of polymer concrete floor coatings, apart from the proper selection of raw materials and their composition, is the thorough preparation of the base. Without optimum adhesion of the coating to the base, there can be no effective covering at all. The surfaces on which the coating is applied should be prepared particularly thoroughly as it helps later to build up clean, jointless coverings which do not lose strength even after prolonged loading.

Before applying the polymer concrete coating, the floor surface should be thoroughly cleaned and all the loose elements on the base removed. This work is carried out using brushes and industrial dust removers or by special machines when the coating has to be applied on large surfaces. To ensure optimum adhesion of polymer concrete to the base, a ground coat in the form of a synthetic resin is applied on the floor surface. The chemical composition of this resin is selected taking into consideration its compatibility with the polymer concrete coating. The polymer coating can be applied only after the application and polymerisation of the ground coat.

Continuous production of the polymer concrete mix is recommended when coating large floor surface areas. Companies carrying out such work have appropriate machinery and equipment which is set up on mobile frames or directly on the chassis of a transport vehicle designed for this purpose. The raw material as well as the power equipment can be easily transported from one target area to another and its mobility right on the construction site can also be ensured. Depending on the width and thickness of the coating and also on the productivity of the equipment, the rate of advance of the transport vehicle is determined. From the end of the chassis, the polymer concrete slurry ready for application falls on the surface being treated, is spread and then compacted. Further, the devices and the technical means used are the same as when handling other similar materials: vibrating brushes, worm conveyors, and spreading and compacting vibration systems.

The most optimal method depends mainly on the thickness of the coating and the consistency of the mass applied. The laying of a thick layer of relatively dry slurry (for example, up to 2 cm) undoubtedly calls for a different approach than when applying a thin (for example, about 4 mm) film of free-flowing slurry.

When applying coatings in industrial or warehouse buildings, the floor should be uniform and smooth. When applying coatings on roads, the surface should be rough for adequate protection of vehicles from slipping on the road. For this purpose, on a surface that has already been coated but the coating has not yet hardened, fairly coarse fractions of the filler are dispersed. These fillers penetrate into the polymer concrete mass only partly. After complete hardening and drying of the solution, the large granules of the filler firmly penetrate into the coating system and are not pulled out by the wheels of vehicles even when the road traffic is fairly intense.

When continuously processing, with appropriate organization, all the individual operations (preparation of the base, applying the ground coat, preparation of the polymer concrete mass, distribution and compaction of the slurry, sprinkling the filler, hardening, and cleaning the finished coating) can be carried out in a continuous manner and so quickly that a given road section can be repaired with traffic moving and blocking only some sections of the road for a short period. Within a very short time (when applying the coating on a small area, this period is roughly 30 min and on relatively large areas a few hours), regular traffic can resume on the renovated section of the road.

Preparation of nonporous polymer solution

As already pointed out, any air present in the casting can be almost completely removed by optimising the size composition of the filler and the composition as well as by good ramming of the mix. However, air cannot be removed completely from the polymer concrete and mould. In many areas of application of polymer concretes, there is no need for this at all but there do exist areas where an absolute non-porous coating or product is required, as for example, in the production of electrical insulators for which non-porosity is prescribed for technical reasons, and also for producing decorative polymer concrete slurries which, after hardening, are subjected to additional cutting and sawing; there should then be no visual differences in the quality of cast surfaces and cut edges.

For complete removal of air from the polymer concrete slurry, it should be subjected to degassing. At present, this measure too can be carried out in a continuous regime; earlier, this used to be effected by filling individual moulds in a vacuum chamber. During continuous processing, the polymer concrete slurry prepared in a casting machine also passes through a vacuum chamber in which air is removed and a non-porous polymer mass is obtained. This mass then enters the open mould in which hardening takes place (Fig. 9.10).

When working in a continuous degassing unit, it should be borne in mind that the presence of the slurry in a thin layer having a correspondingly large area in the vacuum chamber and adequate rarefaction promotes degassing.

When producing a non-porous decorative polymer concrete (for example, imitation marble), it is recommended that the pigment be injected into the solution outside the casting machine in the vacuum chamber, where injectors can operate in the same working regime.

Casting polymer solution under pressure

The expanding effect of air and water on the structural materials necessitates increasingly intense repair of surfaces and structural members using other materials and methods, especially by applying polymer concrete coatings. Polymer concrete coatings are also applied on vertical surfaces, which is only possible by using the appropriate equipment produced by Respecta Company. The machines can process various resin systems (for example, polyurethanes, epoxides, and polyesters) and also various fillers (quartz sand and also light fillers). The latter are of definite importance as insulating material.

The casting machine consists of two sections: the doser and the sprayer. Both sections are connected by hoses up to 50 m in length; a very large doser section can be set up beyond the immediate proximity of the target area and even beyond the limit of the construction site in a more convenient place. The sprayer is designed in such a way that its operation can be controlled by a single person. Both the components of the solution in the form of liquids are fed to the doser from barrels by pumps with electric drives while the filler is fed by a paddle

Fig. 9.10. Casting machine with special degassing unit for producing a dense polymer slurry.

feeder. Dosing of the various materials can be controlled at the required level independent of each other. In the sprayer section, both the liquid components are initially mixed using a static mixer and then (in a different mixer) are mixed with the filler. From the mixing chamber, the finished polymer concrete slurry enters the nozzle and is sprayed on the surface treated. While spraying, the distance between the surface and the nozzle should not be less than 0.8 m. At this distance, the nozzle can be moved to cover an area of about 1 m². When required, the consistency of the solution can be modified directly in the course of spraying.

At the end of the spraying process, the feeding of the solution components is stopped and the washing system connected. From the container, an integral part of the dosing system, a solvent is fed to the sprayer which washes all the components of the machine which came into contact with the resins and the solutions. Additional blowing with compressed air ensures total evacuation of all the sections and elements of the sprayer and eliminates the danger of the material remaining at unexpected places. The sprayer section can be dismantled in the matter of a few seconds.

Practical experience has demonstrated that it is far simpler to apply a slurry using a sprayer rather than a trowel although definite experience in working with a sprayer is needed to ensure the required thickness of the layer applied. When the slurry is of the requisite consistency, it is possible to apply a layer of a few centimetres in thickness on vertical surfaces.

Safety techniques when handling polymer concrete

All workers handling polymer concrete (and polymer materials in general) must clearly understand that they deal with chemical reagents which to some extent or the other can be deleterious to their health. Therefore, when handling these materials, some measures of preventive care have to be exercised.

The work should be organised in such a way that direct contact between the worker and the material is reduced to a minimum. Special care and vigilance should be exercised when working with continuous casting machines since all the constituents are processed in these machines in closed sections and emerge from the machine as finished polymer concrete mix. This system causes a relatively low atmospheric contamination with solvent vapours. In the course of washing the equipment, which requires an insignificant amount of solvent, the liberation of solvent vapours too remains insignificant and facilitates their suction. Preferably, solvent vapours should be sucked out immediately on their emergence from the casting machine. When using an exhaust system, the physical and chemical properties of the vapours generated should be clearly understood. For example, the styrene vapours arising when processing polyesters are heavier than air and hence their suction is best carried out at a lower level. This can prevent the access of vapours to the head level of the worker operating the machine.

Efficient suction is particularly necessary when applying the thin film layer ('gel coat') on the mould surface using air since, during this operation, very intense liberation of solvent vapours is noticed. For applying the 'gel coat', it is desirable to have a special spraying booth or at least an exhaust hood.

In addition to the normal ventilation systems with which production buildings are equipped, it is recommended that exhaust devices be added at points of intense dust formation, especially during complete or partial processing of the fillers. Dust is dangerous to man's health and can adversely affect the quality of the products produced (especially decorative finishes whose surfaces should be clean and beautiful). Dust settling on a surface to which the 'gel coat' has been applied will later appear as a contaminated surface on the finished product.

Effective suction of deleterious vapours and adequate inflow of fresh air are extremely important to ensure safety when working with these materials and to minimise the danger of spontaneous combustion of mixtures of solvent vapours and air. Special attention should be paid to these aspects when processing resins having a low ignition point.

Since it is not always possible to avoid direct physical contact between the men and the resins and solvents, it is necessary to understand that the solvents intensely desiccate the skin. Hands, which are particularly exposed to this danger, should be regularly treated with a fat-base cream coating. A good impregnation of the skin with the cream prevents the adhesion of the resin and eliminates washing the hands with a solvent.

The consumer should obtain from the manufacturer a complete description of the chemical reagents used and conform to all the stipulations. The descriptions contain, for example, certain standards and instructions and are based on the appropriate instructions prevailing in every country. It is possible that such instructions are interpreted differently in different countries. Some essential general provisions are outlined below.

An important criterion for the classification of resins is the ignition point. If this index falls below the normal room temperature, preparation of the polymer concrete products should be exclusively carried out in covered buildings and the entire equipment should be explosion-proof, i.e., all the electrical equipment (machinery, illumination, and switches) should be so designed as to prevent the generation of sparks that might ignite the mixture of solvent vapours with the air. In this respect, methacrylate resin, with an ignition point at around $21°C$, is particularly hazardous.

Special care should be exercised when working with activators, especially when using peroxide or acid hardeners. Many operators do not wear protective goggles as they come in the way of their normal working. Invariably, a bottle of water should be available for washing the eyes when urgently required. Peroxides supplied in cylinders filled under pressure should not be used since hermetically sealed cylinders on sudden rarefaction of peroxides can act like a bomb. This may result in a sharp increase in the volume of the contents in the cylinder. To minimise the danger of rarefaction, which is uncontrollable and sudden, peroxides should be stored in a special building away from the production building and also from the site of storage of accelerators.

When selecting the type of machinery and equipment for processing the raw material, the working pressure at which the machines are operated should also be taken into consideration. A low working pressure is more reliable and hence machines which operate at rather low pressures should be selected.

Nearly all chemical reagents ignite. Accordingly, appropriate instructions have been worked out for each raw material. All the instructions must be unfailingly fulfilled. Their importance should not be underrated but there is no need for being overly cautious. Gasoline pumps which are handled by opera-

192

tors sometimes with a lighted cigarette dangling from the mouth, are far more dangerous than workshops producing polymer concrete since there is a large quantity of gasoline in the filling station. Nevertheless, smoking in general is totally prohibited in workshops producing polymer concrete products.

Technological equipment

Fig. 9.11 depicts a technological flow sheet for producing drainage pipes, feed hoppers, columns and other products which are quite popular among the polymer concrete products produced.

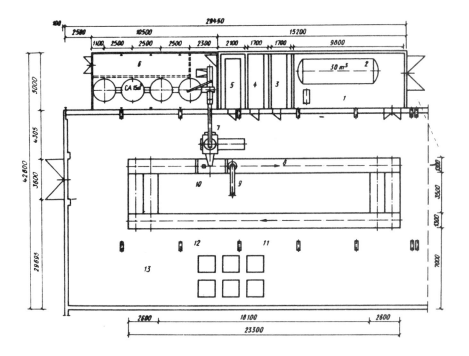

Fig. 9.11. Technological flow sheet for producing engineering products using polymer concrete:

1—storage for polyester resins; 2—smaller storages; 3—accelerators; 4—hardeners; 5—service building for electrical equipment; 6—system for filler preparation; 7—casting machine; 8—roller conveyor; 9—exhaust; 10—filling and compacting mass in the moulds; 11—striking; 12—applying the separating layer on the mould; 13—storage of finished products.

The size composition of the filler comprises four different fractions and the composite mixture later goes into the casting machine. The liquid materials (resins, catalysts, and accelerators) are fed into the casting machine through flexible pipe lines (or hoses). All the storage buildings are located outside the production buildings.

The polymer concrete mass is prepared in the casting machine and fed from the machine directly into the moulds brought on a roller conveyor under the outlet of the casting machine. While filling the mould, the polymer concrete mass is compacted by vibrations. After 10–15 min, the moulds move towards the other end of the roller conveyor where the finished products are removed from them. The finished products go to the storage and the moulds are returned for cleaning. Later, a new separating layer is applied on them and they are returned to the casting machine for a fresh filling of the polymer concrete mass.

Usually, when producing engineering goods, a 'gel coat' is not applied on the moulds.

A unit for producing decorative panels is shown in Fig. 9.12.

The size composition comprises two different fractions (sand and flour) and after adding pigment (powder), the mix is charged into the bunker of the casting machine. Liquid constituents are stored in containers outside the production shop and are fed through pipe lines to the casting machine. Pigments required for producing artificial marble or imitating the surface of natural marble, are prepared close to the casting machine and later charged into the receiver-storage of the casting machine.

The moulds for preparing panels move on a roller conveyor and come under the outlet of the casting machine and are filled with the polymer concrete mass (technological point A) (Fig. 9.12). The mass is later compacted (technological point B) and thickens in the course of transport to the hardening section (C). Thus, on entering the rotation point (D), the polymer concrete mass is sufficiently stable and does not drop off from the mould as it is rotated. Subsequent movement is done in trolleys (chassis on wheels) in which the moulds filled with the mass are laid on several shelves one over the other and the polymer marble is allowed to harden completely before striking. In this period, the trolleys with the trays stand in the hardening section (C).

After sometime, the trolleys are moved to the technological point (E) for removing the finished panels from the moulds. Here, the moulds are removed together with the finished products and the polymer marble panels remain on the shelves and are transported to auxiliary buildings for additional hardening or processing (cutting, polishing, etc.). At technological point F, the moulds are cleaned, a new separating layer is laid and polished at technological point G to ensure lustre of the panel surface. The moulds then move on roller conveyors to point H where the 'gel coat' is applied by spraying. The aerosol formed during spraying of the material is simultaneously removed by exhaust fans. The moulds with the thin film coating are returned to the trolleys where they remain until the thin film layer has hardened. Later, the moulds are again laid on the roller conveyors which transport them to the casting machine for filling again with the polymer concrete slurry.

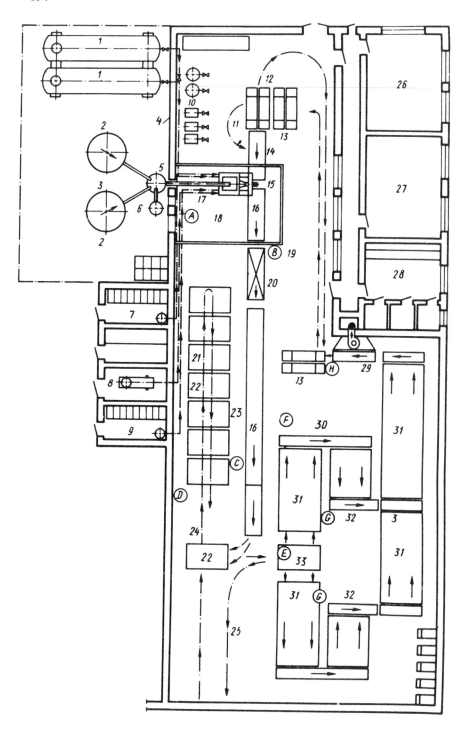

The use of trolleys is highly recommended in the production of flat large-sized structural products such as panels and slabs. This makes for optimum utilisation of the production area.

Figure 9.13 depicts the technological flow sheet for producing panels and slabs as well as various sanitary-engineering products. The system of storage and supply of raw material has been designed on the principles detailed in the preceding flow sheets.

After processing the polymer concrete slurry in the casting machine, the slurry is fed directly into the moulds which are brought to the casting machine on a roller conveyor and after filling with the polymer concrete mass move for further processing in the technological cycle. When using identical formulations, it is not important what is being produced—be it panels or sanitary-engineering equipment. After compaction, the moulds move again on the roller conveyor but are separated according to the types of products. The moulds with panels are laid in stacks on trolleys and are subjected to further processing (see Fig. 9.6) while the sanitary-engineering articles are graded according to types and are transported on roller conveyors branching from the main conveyor (perpendicular to it). At the end of a short period, the outer moulds forming the backside of products (these moulds are not covered with the thin film layer) are removed and sent for storage. On their way to storage, a separating layer is applied on the moulds. Later, the outer moulds remain in storage until the corresponding internal mould is received.

Soon after removing the outer moulds, the inner moulds with the finished sanitary-engineering products are also removed. The striking of large-sized articles (tanks, washing machine shells, bathtubs, etc.) is done using striking frames. Large-sized products are taken out from the production buildings on supporting frames for thorough hardening. After hardening, additional operations for bring-

---

Fig. 9.12. Technological flow sheet for producing decorative panels:

1—containers for storing polyester resins; 2—fillers; 3—bunkers; 4—pipe-line for transporting the polyester resin; 5—hopper; 6—base pigment; 7—catalysts; 8—compressed air; 9—accelerators; 10—technological point of mixing; 11—additional pigments; 12—technological point of hardening the thin film layer ('gel coat'); 13—mould trolleys; 14—moulds ready for filling; 15—outlet channel; 16—roller conveyor; 17—worm conveyor; 18—technological point of mixing and dosing; 19—technological point of compacting the mass; 20–vibration table; 21—pipe lines connected to the mixing unit; 22—trolley storage; 23—hardening section; 24—technological rotation point; 25—withdrawal of trolleys for further hardening of the mass; 26—service building; 27—workshop; 28—storage building; 29—technological point for applying the thin film layer; 30—technological point for cleaning and applying the separating layer; 31—empty moulds; 32—technological point for polishing; 33—technological point for striking.

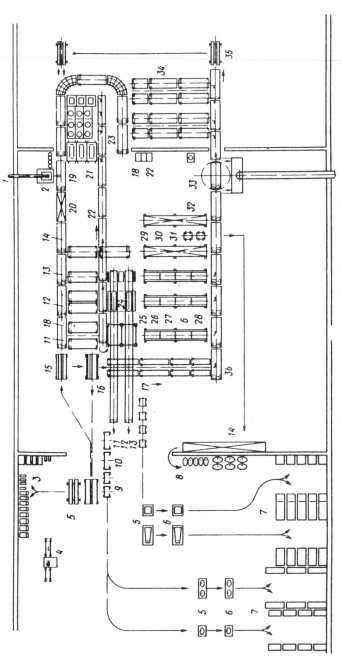

Fig. 9.13. Combination unit Respecta for producing panels and sanitary-engineering products:

1—system of producing the filler; 2—casting machine of Respecta Co.; 3—storage of panels; 4—saw for cutting the panels; 5—grinding of the products; 6—polishing of the products; 7—storage; 8—storage for small products; 9—supporting frames; 10—washing devices; 11—service tubs; 12—washing tables; 13—shower cabins; 14—grinding, polishing and bonding of small products; 15—trays for transporting the panels; 16—striking of panels and slabs; 17—moulds for producing panels; 18—panels; 19—filling the moulds with polymer concrete mass; 20—compaction of the mass; 21—storage of outer moulds; 22—transport of outer moulds; 23—assembling the moulds; 24—striking; 25—internal moulds; 26—cleaning and applying the wax; 27—polishing; 28—separating lacquer; 29—striking the small-sized products; 30—applying the wax or paste on the mould followed by polishing; 31—casting; 32—applying the thin film layer ('gel coat') on small-sized articles; 33—spraying the thin film layer on the moulds; 34—hardening of the thin film layer; 35—trolleys for transporting panel moulds.

ing the product to the required condition (grinding, polishing, etc.) are carried out. Much smaller products (soap holders, mirror frames, and various types of toilet room fixtures) are removed from the mould on the striking table and are left there for thorough hardening. They are then processed for bringing them to the required condition (as and when required).

The inner moulds of sanitary-engineering products are cleaned after striking and covered with a separating layer. A priming paste is used as a separator and polished before applying the separating lacquer (polyvinyl acetate) which itself forms a separating layer. After hardening of the separating layer, a thin film layer is laid on the mould surface. For this purpose, provision is made in the technological flow sheet for a rotating table with a system of spraying the material protected by an exhaust screen. After hardening of the thin film layer, the inner and outer components of the moulds are again assembled and returned to the casting machine for the next cycle of moulding.

9.7. Preparation of Polymer Concrete Products and Structures

The integrated experimental and theoretical research extensively carried out in the Soviet Union in the field of physical and chemical principles of structure formation and rational technology of highly efficient polymer concretes, including research on the morphology of supramolecular formations relative to composition and temperature, the study of shrinkage stresses using models of adhesive compositions (i.e., microstructure and macrostructure of polymer concrete), basic parameters of preparation, vibration moulding, and thermal treatment, have resulted in developing optimal compositions of polymer concretes and the plant technology necessary for producing them.

At present, in the various branches of the industry, there are dozens of units producing polymer concrete products and structures. Thus, in Dzhezkazgan, polymer concrete blocks, columns, foundation blocks and other structures are produced. In Svetlogorsk, foundations for pumps, roofing panels and floor panels are produced. The production of decorative polymer concrete panels has been organised in Ashkhabad and Tallin. There are two units each producing heat- and chemical-resistant polymer concrete panels for floors of livestock buildings in the Krasnodarsk region. The commercial production of polymer concrete pipes for mine timbering has been perfected in Prokop'evsk while the commercial production of electrolysis baths and other chemically stable structures has been perfected at Ust'-Kamenogorsk, Noralsk and other towns.

The shop for polymer concrete production at Ust'-Kamenogorsk has a capacity of about 1500 m^3 a year of various chemical-resistant reinforced polymer concrete structures, primarily electrolysis baths. Moreover, this shop has commissioned an automated line for producing glass-plastic reinforcements with an annual capacity of 10 tonnes [30].

The polymer concrete structures shop consists of the following divisions:

— Crushing and grading, I
— Drying and storage, II
— Production of polymer concrete structures, III
— Production of glass-plastic reinforcements, IV
— Storage of materials and finished products (Fig. 9.14).

The coarse filler (quartz and andesite) enters the crushing-grinding section where it is subjected to two-stage crushing, initially in jaw crusher 1, and later taken to roll crusher 3 by elevator 2 for final crushing to the required size. The resultant gravel fraction-wise goes by conveyor 4 into bunker 5 of the drying-storage section. Steam dryers are provided in the bunkers for drying the fillers and aggregates. When required, the aggregates are additionally dried in rotating drum furnace 6.

The fillers and aggregates from the drying-stroage section go according to size fraction by elevator 8 to feeder 7 and enter bunkers 9 and 10 (for gravel of two fractions) and 11 and 12 (for sand and andesite flour). The FAM resin (furan type) is charged into storage container 13, benzenesulphonic acid into bath 14 with a water jacket for melting it. The gravel, sands, and andesite flour are fed into concrete mixer 17 by dosers 15 and 16. The FAM resin and benzenesulphonic acid enter mixer 20 through dosers 18 and 19 for mixing for 15 to 20 sec. The binder obtained is then fed to concrete mixer 17 where it is mixed with the fillers and aggregates.

The shell and the glass-plastic reinforcement are assembled on stand 21. Later, shell 22 is placed on trolley 23 and brought under concrete mixer 17 for charging the polymer concrete mix, after which an overhead crane places it on vibration table 24 for vibration compaction. For a better quality compaction by vibration, large-sized articles like electrolysis baths are provided on the side walls with suspended vibrators. The moulded products are placed in heating chamber 25 in which thermal treatment is carried out.

The glass-plastic reinforcement is produced on an automated line which consists of bobbin holder 26, unit for impregnation and thermal treatment 27, unit for winding the finished product 28 and control panel 29.

However, these works, including the units at Ust'-Kamenogorsk, do not meet the current demands. The Giprotsvetmet Institute together with the Reinforced Concrete Research Institute developed a design for highly mechanised works with a conveyor system for producing wide-ranging polymer concrete products and structures for industry, agriculture, civil, and other branches of the building industry.

In developing the technology and design of such units, the experience gained in the existing shops at Ust'-Kamenogorsk, Dzhezkazgan, Moscow Copper Elec-trolysis Works, and also the experience of producing polymer concrete structures at the Almalyk and Balkhash Mining-Metallurgical Combine was utilised.

The specific features of the developed and adopted technology are: possi-bility of producing polymer concrete products of uniform structure and constant

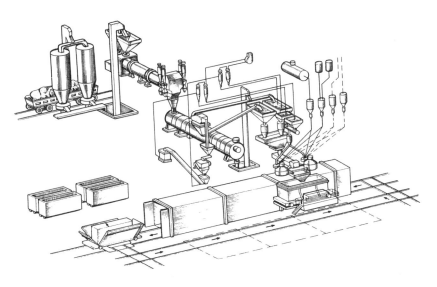

Fig. 9.14. Production of reinforced polymer concrete electrolysis baths. Sketch depicts production area and that section of the shop producing the glass-plastic reinforcements. [Numbering of apparatus (as explained in text) missing in Figure—Translator]

physical and mechanical properties; periodic or continuous technological process as required by the overall sizes of the finished products and type of binder (operations of vibration moulding, thermal treatment and application of the protective coating are carried out on a moving ground-level conveyor while thermal treatment of the products is carried out by an accelerated regime in special aerodynamic heating furnaces, type PAP); standardisation (the same conveyor can be used to produce components of different lengths, widths and heights, compact, ribbed, or spatial, using heavy or light polymer concrete based on claydite or agloporite aggregate with maximum dimensions of products at 12,000 mm length, 3000 mm width and 3000 mm height; changes in the nomenclature of the products are due to slight modification of the moulds); possibility of producing large-sized articles with a high degree of finish which reduces the expense of labour in construction and the duration of raising the building or installation; ensuring the given output of 5000 to 10,000 m^3 a year depending on the number and productivity of the selected equipment.

The developed technology of producing polymer concretes helps fabricate structures based on furan resins type FA or FAM, polyester type PN-1, carbamide, methyl methacrylate, etc. [30].

The technology of producing polymer concrete articles consists of four main stages: preparation of the constituents, preparation of the polymer concrete mix, moulding and vibration compaction of the products and their thermal treatment (Fig. 9.15).

The dried fillers and aggregates and polymer binder with the help of dosers 4 enter concrete mixer 5.

The polymer concrete mix is compacted on a resonance-type vibration platform with horizontally controlled vibrations. The amplitude of vibrations is 0.4–0.9 mm in the horizontal and 0.2–0.4 mm in the vertical direction and the frequency of vibrations 2600/min. The duration of vibration compaction is two minutes.

The laying and vibration compaction of the mix are carried out in a closed building equipped with suction and exhaust ventilation. Simultaneous with the moulding of the polymer concrete structure, control samples 100 mm × 100 mm × 100 mm are also moulded for determining the grade of strength of polymer concrete. Three such control samples are produced for every article made of polymer concrete of volume 1.5 to 2.4 m^3.

To produce products with given properties in a much shorter period, they are transported on a ground-level conveyor to the thermal treatment chamber. Thermal treatment is carried out in an aerodynamic heating furnace, type PAP, which ensures uniform temperature distribution throughout the body.

At the end of thermal treatment, the finished products are automatically transported by a conveyor into a process building where the products are removed from the mould and sent to the product storage while the empty moulds are

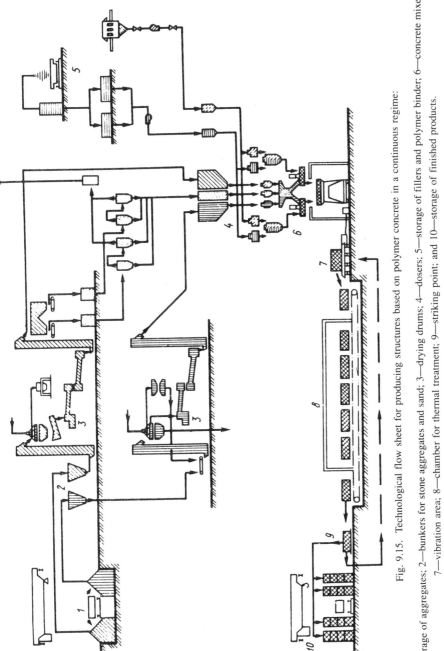

Fig. 9.15. Technological flow sheet for producing structures based on polymer concrete in a continuous regime:

1—storage of aggregates; 2—bunkers for stone aggregates and sand; 3—drying drums; 4—dosers; 5—storage of fillers and polymer binder; 6—concrete mixer; 7—vibration area; 8—chamber for thermal treatment; 9—striking point; and 10—storage of finished products.

cleaned of extraneous matter and remnants of polymer concrete and prepared for the next cycle of moulding.

The entire plant is serviced by twenty persons working in two shifts. All the operations are mechanised and automated to the maximum extent conforming to the adopted technology. In the technological process, powerful suction and exhaust ventilation systems are provided to ensure normal working conditions in the shop. The exhaust gases are burnt in a special unit. The process effluents go to the cleaning unit.

Apart from the main technology used for producing polymer concrete products and structures, the shop can be provided with divisions for producing glass-plastic reinforcement and preparation of polymer solution based on FAM resin in the form of two packings: dry mixture of microfiller, sand and hardener, and resins with modifying additives. The polymer solution from these two packings is produced directly in the constuction area by mixing. In the shops of the Ministry of Non-ferrous Metallurgy, provision has been made for divisions to produce special ampoules for timbering mine workings.

The developed design has begun being used at four highly mechanised factories in the field of non-ferrous metallurgy and other branches of the industry, including the fabrication of a shop of capacity 5000 m^3 of polymer concrete structures, 2000 tonnes of the two-component solution, 20 tonnes and 10,000 [sic] ampoules a year of glass-plastic reinforcement. The design of two mechanised units has commenced with a capacity of 10,000 to 15000 m^3 a year for Ozer and Udokan Mining-Concentration Combines.

The large-scale experimental-commercial introduction of electrolysis baths for zinc and copper electrolysis at five plants of non-ferrous metallurgy which have been in operation for up to 15 years has demonstrated the reliability of the material used for making the containers.

In the Soviet Union, interest has recently been evinced in artificial structural finishing materials. The Reinforced Concrete Research Institute, Institute of Seismic-Resistant Structures, Ashkhabad and other organisations are engaged in developing compositions and plant technology for producing surfacing materials from polymer slurries and polymer concretes based on polyester resins, methyl methacrylate, and other binders.

Much interest is shown in this problem in regions where marble and other natural finishing materials are not available. This problem is particularly acute in the Turkmenian Soviet Socialist Republic which has almost unlimited reserves of Karakum barhan (sand-dune type) sands and extremely limited deposits of natural building materials. Therefore, one of the first plants for producing polymer concrete finishing panels was set up at the Ashkhabad Combine of building materials and structures. The developed compositions use over 50% barhan sands and this type of polymer concrete has been given the trade name 'Barkhanlite'.

'Barkhanlite' is a fine-grained polymer concrete consisting of local fillers, aggregates and a synthetic binder.

A characteristic feature of the developed technology of Barkhanlite production is the possibility of producing articles with a uniform structure, constant physical and mechanical properties, wide colour range, and cyclic or continuous technological process. Moulding, vibration compaction and thermal treatment of the articles are carried out on a moving conveyor belt. Thermal treatment of the products by an accelerated regime is carried out in special polymerisation chambers. The flow line can be used to produce components of different sizes and shapes from heavy or light polymer concrete. The maximum dimensions of the products are: 300 cm × 80 cm × 2 cm. A change in the sizes of finished articles involves only an insignificant expense in modifying the mould.

The capacity of the plant is 25000 to 50000 m^3 of products a year depending on the productivity of the equipment used. Polyester resins and methyl methacrylate are used as synthetic binders for Barkhanlite. Conforming to the production technology, the plant consists of four main sections; preparation of the components, preparation of the Barkhanlite mixture, moulding, compaction of products and thermal treatment.

The constituents of 'Barkhanlite', i.e., washed mineral matter and sand, are brought by railroad or automobile transport systems and loaded in the receiver bunkers; the inert matter is fed separately into a drying drum on a conveyor belt. The temperature of the material at the exit of the drum is 100–110°C. From the drying drum, the aggregates go to the sieves on an elevator. After sieving, the inert material enters the storage bunker. From this storage bunker, the aggregate goes on a bucket elevator according to sizes to another bunker. The microfiller (ground barhan sand) after grinding in a ball mill also goes on an elevator to the bunker. The dry inert constituents are weighed in quantitative dosers and go on a belt conveyor and elevator to a 250-litre induced mixer for preliminary mixing of the dry constituents. After mixing, the dry constituents go to the bunker of the mixing unit on a belt conveyor.

Polyester resin PN-1 or any other binder as also the hardener are pumped from the material storage unit through a pipe line into containers for the resin, initiator of polymerisation and hardening accelerators.

The polymer concrete mix is produced in the mixing unit of Respecta Co. Before being fed into the mixer system, the polymer binder (PN-1 or PN-3) is heated to 50–60°C in the same unit. The polymer concrete mix goes in the finished form directly from the mixing system of the unit.

The surface of the panel is levelled manually using a special fluoroplastic putty knife and vibrated directly in the mould on a horizontal vibration table at a frequency of 2500 to 3000 vibrations/min and amplitude 0.5 mm/15–20 sec.

The top surface finish layer of the facing panels is laid before the mix is moulded at a special point by laying the finish composition at the bottom of the mould followed by levelling it with a fluoroplastic putty knife and partial hardening in a continuous chamber at $50 \pm 10°C$ for 25–30 min. After this, the mould with the finish layer goes for moulding on a roller conveyor.

The filled-in moulds with the compacted mixes go to the polymerisation chamber on roller conveyors on which they move continuously for 45–60 min. The temperature in the polymerisation chamber for panels based on polyester resins is 80 to 100°C and for panels based on methyl methacrylate $30 \pm 10°C$.

From the polymerisation chamber, the moulds with the panels go for striking. Later, the panels are graded and sent for marking, technical control, packing and storage of finished products. The moulds, after removal of the products, are returned for preparation, cleaned and lubricated, and are then sent to the chamber for moulding. The moulds received for moulding are heated to 30–40°C for accelerated polymerisation of the polymer concrete mix.

A distinctive feature of the technology of producing Barkhanlite panels is the moulding of the polymer concrete mix in horizontal moulds with glass or metal base polished to the seventh class. The specially developed separating lubricant together with such bases helps produce a facing surface on the panel that requires no polishing or grinding.

The entire shop is operated by 12 persons in one shift. Conforming to the adopted technology, all the operations are mechanised and automated to the maximum extent. Suction and exhaust ventilation ensures normal working conditions in the plant. The plant capacity is $25000 \, m^3$ of products a year.

A similar plant using a mixing unit produced by Respecta Co. has been set up at Dezintegrator Combine in Tallin. Unlike the Ashkhabad plant, only polyester resin is used here and the moulds are fully metallic.

Fig. 9.16 shows the technological flow sheet of a shop for producing polymer concrete products based on carbamide resins, phosphogypsum and lignin [84].

Phosphogypsum is transported and loaded into receiving bunker 1 from where a conveyor takes it to gypsum curing pan 2 and later the dehydrated phosphogypsum goes to doser 8.

Lignin is transported and loaded in receiver bunker 4 and then a conveyor takes it to drying drum 5. After drying, the lignin is sieved into three fractions and stored in bunker 7.

The dosed amount of phosphogypsum and lignin are charged into concrete mixer 9 in which the dry mix is mixed. Later, carbamide resin is added to the concrete mixer from container 3 through doser 8 and the mixture agitated until a uniform mass is obtained.

The prepared mix is cast in mould 10, compacted by vibration, and the product together with the mould sent to thermal treatment chamber 11. Thermal treatment is carried out at 60–70°C for 3–4 h. After thermal treatment and striking, the articles are sent to the finished products storage unit.

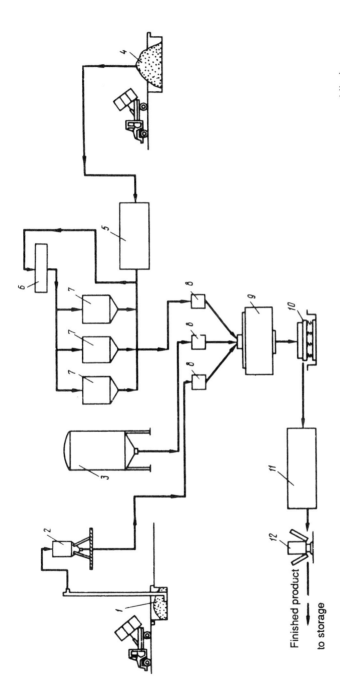

Fig. 9.16. Technological flow sheet for producing polymer concrete articles based on carbamide resins, phosphogypsum, and lignin:

1—phosphogysum storage; 2—calcination of phosphogypsum; 3—container for carbamide resin; 4—lignin storage; 5—lignin drying; 6—lignin sieving; 7—intermediate bunker for lignin; 8—dosers; 9—concrete mixer; 10—vibration area; 11—thermal processing chamber; 12—striking.

Finished product
to storage

206

9.8. Prospects for Growth of Technology of Polymer Concrete Products and Structures

Experience in producing polymer concrete mixes in various countries using standard concrete mixers designed for producing cement concrete has revealed several vital drawbacks in using such equipment: high resin consumption, inadequate mixing quality and periodicity of mix preparation.

Highly effective equipment and technology have been developed in some industrially advanced countries. Thus, the Italian firms Snia-Wiscosa, Longinotti, and Breton developed technological processes for producing decorative facing panels of polymer concretes using process wastes of marble and other natural rocks as fillers and aggregates and polyester and acrylic resins as binder. The technological process and the equipment used by the firm Breton are most interesting.

The principle of the Breton technological process is as follows: the polymer concrete mix is mixed in a special unit (Fig. 9.17), charged into a mould and compacted by vibration; later, the mould goes to a powered system in which it is subjected to vibration pressing. All of the processes of mixing, charging, vibration moulding and vibration pressing are carried out under vacuum. The blocks produced are cut by diamond saws into panels of required sizes.

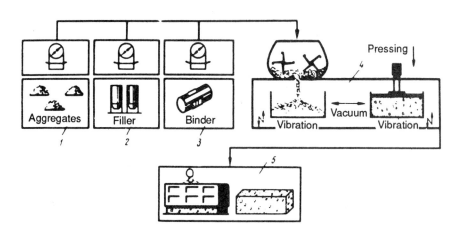

Fig. 9.17. Main flow sheet for producing defect-free blocks of volume up to 3 m³:

1—storage of aggregates; 2—storage of fillers; 3—container for polymer binder; 4—vacuum chamber for mixing under vacuum, vibration compaction, and vibration pressing; 5—striking.

This technology yields polymer concrete blocks of 1 to 3 m³ without air inclusions and with the maximum possible density. Thus, blocks totally free of defects all through the section are produced. The capacity of the unit is up to 5000 m² a day [*sic*].

Among the drawbacks of this unit are the comparatively high metal and power consumption and hence the high cost of the equipment. Moreover, appropriate equipment is required for sawing the blocks, thus involving additional expense.

The Japanese company Nippon Telegraph and Telephone (NTT) developed an extremely interesting tunnelling system for laying telephone cables. In this system, it is possible to make underground tunnels of inner diameter 1200 mm, wall thickness 140 mm and length up to 500 m at 10 m a day. This system is distinguished by the method of producing tunnel walls using fine-grained polymer concrete based on polyester resins (Fig. 9.18). The wall made in the advancing portion of the tunnel should withstand the thrust of the cutting head of the tunnelling machine. In order to ensure the above rate of driving, the polymer concrete should set and provide adequate strength in 30–40 min.

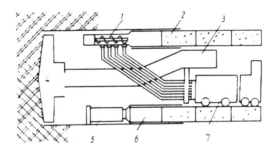

Fig. 9.18. Main flow sheet for driving and laying the tunnel walls:

1—mixer; 2—tunnel wall; 3—conveyor; 4—cutting machine; 5—jack; 6—mould; 7—container with polyester resin.

To ensure rapid setting and to obtain the required strength, the company uses a fine-grained polymer concrete composition with a binder based on unsaturated polyester resin 'Rigolac' produced by the firm Seva Kobunei and hardeners cobalt octanate, methyl ethyl ketone, as also aromatic and aliphatic co-polymers. The binder content in the polymer concrete is 20% by weight. Dried river sand of grain size up to 3 mm and calcium carbonate powder of particle size 0.5–25 μm are used as fillers.

The bending strength of the polymer concrete without the co-polymer after 2 h of hardening is 20–22 MPa; it is about 30 MPa in compositions containing co-polymers after 50–60 min.

The need often arises in building practice for laying protective coatings based on polymer plasters. The conventional production and laying of polymer plaster consists in dosing, mixing (often manually), and applying it with a putty knife (also manually) on the surface to be protected. In this method, quality control of the operations is complicated since everything depends on the experience of the worker. The Japanese firm Toisei Corporation designed a unit which totally mechanises the production and application of polymer plasters on horizontal, vertical and ceiling surfaces.

Epoxide resin is used as a binder and aliphatic polyamines as a hardener. The ratio of filler to resin may vary in the range 3.5:1 to 6.0:1. The unit has been successfully used for applying coatings on tunnels and walls of industrial sewer canals.

An analysis of the experience gained in large-scale production of polymer concrete and reinforced polymer concrete products and structures in various countries has helped to identify the main directions in the development of the contemporary technology of producing such products and structures and to lay the course for evolving the third generation of industrial establishments for producing them.

Firstly, investigations will continue on using new types of monomers and oligomers, catalysts and accelerators, including those possessing very high strength characteristics, chemical and thermal stability, and other favourable properties. Different types of superplasticisers will be used in much greater volumes to reduce the duration of vibration moulding while moulding by casting will be adopted in several cases. The use of new types of hardeners and superplasticisers will help to control the durability of polymer concrete mixes in a wide range, reduce the consumption of the polymer binder, result in a very dense structure and, correspondingly, ensure very high physical and mechanical characteristics.

The successful experience of using continuously operating units of the type Respecta for producing polymer concrete mixes has confirmed that the polymer concrete mixes produced using polyester, carbamide, phenol-formaldehyde, and other resins are of a better quality. The computerised operation of such units helps in achieving a qualitatively higher level of production of polymer concrete mixes.

Vibration moulding of reinforced polymer concrete products and structures is presently one of the main methods which usually employs standard vibration platforms with frequencies of 50 Hz and amplitude 0.3–0.5 mm. Since most synthetic resins possess several specific properties, i.e., high viscosity (many times exceeding the viscosity of water used as a binder in cement), considerable stickiness and adhesion strength, the frequency, amplitude, duration and direction of vibrations most frequently used in vibration moulding of cement concrete are less effective when moulding polymer concrete products, even when employing standard vibration platforms. In many cases, overloads have to be applied and the duration of vibration moulding extended, thus complicating the technology and consuming more power.

According to Ur'ev and Mikhailov [79], the rate and shear stresses in highly filled polymer compositions should correspond to the least value of effective viscosity for quality compaction in highly filled polymer compositions. The authors point out that from the viewpoint of physical and chemical mechanics, the most effective method of generating such conditions is by high-frequency vibration moulding at frequencies of 10,000 vibrations/min and amplitude 0.2 mm. However, development of the precast reinforced concrete industry shows that the overall tendency of vibration moulding is towards low-frequency vibration moulding on vibration platforms with vibration frequencies of 600–900/min and amplitude 3–10 mm.

Investigations on low-frequency vibration moulding of highly filled compositions showed that such moulding is quite effective and should be more extensively adopted in new shops and plants for producing reinforced polymer concrete structures.

A detailed review and analysis of the various methods of hardening of polymer concretes [62] revealed that the hardening of polymer concretes under normal conditions at 18–20°C for 28–30 days, except for polymer concretes based on MMA (methyl methacrylate), cannot ensure the maximum possible polymerisation of the polymer binder. The heating of structures or articles for 6 to 10 h at 60–70°C after a day's retention under normal conditions also does not ensure the required total hardening.

The originally proposed method, i.e., a day's hardening at 18–20°C and dry heating at 80°C for 20–24 h, helped achieve the maximum possible degree of polymerisation for a wide nomenclature of reinforced polymer concrete products and structures. This method is used in many existing works but suffers from vital drawbacks: the total hardening duration is 44–48 h and power consumption is high. These aspects greatly complicate the technological process and increase the cost of polymer concrete structures.

On the basis of several investigations, a second method of thermal treatment of large-sized products was adopted; on completion of moulding, the polymer concrete articles are held in the mould at 18–20°C for 1.5–2 h. By this time, the heat liberated during the exothermal polymerisation reaction of the polymer binder raises the temperature of the polymer concrete mix to 60–70°C. The heated article together with the mould is placed in a thermal treatment chamber in which the temperature is raised to 80°C. The article is held at this temperature for 16–18 h, after which the temperature is gradually reduced to 20–25°C over 2–3 h (Fig. 9.19).

As a result, the total hardening time is reduced by more than half and power consumption in heating the hardening articles by placing them in a thermal treatment chamber after a day's retention is greatly reduced compared to the earlier practice. In this method, the heat of the reaction is utilised more rationally and the completeness of hardening is practically no different from the indices characteristic of the hardening regimes in vogue.

This method of hardening has already found practical application at one of the operating plants and will be used in future. A necessary condition for adopting this method is that the weight of the polymer concrete in a given mould should not be less than 300–350 kg.

For thin-walled structures with a large heat liberation surface and for structures with a small weight of polymer concrete, this method is not inapplicable but is less effective.

That power economy at all stages of production is presently of paramount importance needs no further justification or proof.

In the production of polymer concretes, thermal processing is one of the most power-intensive processes. Transition to thermal treatment using the heat liberated by the polymer concrete mix reduces power consumption by 25–30%. However, the potential possibilities of polymer concrete mixes have not been exhausted and the development of a hardening process which will help totally to eliminate thermal treatment in special chambers while maintaining all the required characteristics of polymer concretes holds great importance.

It is known that thermosetting synthetic resins liberate in the course of hardening, depending on the type of the resin, 250–300 to 420–480 kJ per g of unfilled resin or 60,000 to 140,000 kJ per m^3 of heavy polymer concrete. It is known that the self-heating of cement concretes is extended and proceeds steadily over several days, which renders difficult the use of the thermos method for hardening cement concretes. In the case of polymer concretes, the polymerisation or polycondensation reaction of the polymer binder proceeds very intensely and the self-heating time is 1.5–2 h (Fig. 9.20).

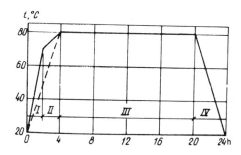

Fig. 9.19. Thermal treatment of polymer concrete structures utilising heat of the reaction:

I—zone of self-heating; II—zone of temperature rise to 80° C; III—zone of retention; IV—zone of cooling.

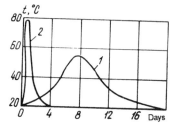

Fig. 9.20. Kinetics of heat liberation:

1—for cement concretes; 2—for polymer concretes.

This pattern of self-heating kinetics of polymer concrete mixes and the significant quantum of heat liberated enable efficient utilisation of the thermos method for hardening polymer concrete products and structures [61].

Experiments have shown that during the hardening of polymer concretes based on FAM [furfural acetone] and PN-1 [polyester] resins of volumes 0.15–0.2 m³ in a mould insulated with phenol foam-plastic of 100 mm thickness, the temperature of the polymer concrete mix rose to 90–100°C as a result of self-heating and was maintained at that level for more than 24 h. When utilising the thermos method in moulding and hardening articles of a volume exceeding 0.2 m³, the self-heating temperature may exceed 100°C. At such a self-heating temperature, thermal fissures may develop in the product.

To eliminate the possibility of thermal fissures, the following method of hardening using the thermos method is suggested. An article of volume exceeding 0.2 m³, is moulded in an ordinary metal mould and held in it for 1.5–2 h. By this time, the processes of exothermal reactions of the polymer binder have ceased and the mix is heated to the maximum possible temperature for a given type and weight of polymer concrete under conditions of heat liberation into the ambient atmosphere. Later, the mould is placed in a thermally insulated tray, covered by a thermos lid (thermally insulated shell), and held in the thermos for 16 to 18 h. Later, the lid is removed and the article allowed to cool to 20–25°C (Fig. 9.21).

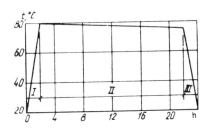

Fig. 9.21. Thermal processing of polymer concrete structures by the thermos method:

I—zone of self-heating; II—zone of retention; III—zone of cooling.

The physical and mechanical properties of polymer concretes hardened by the thermos method are practically no different from the corresponding properties of polymer concretes that have undergone thermal treatment by the above-described regimes. The adoption of this method of hardening would greatly reduce the cost of polymer concrete structures, reduce power consumption and also reduce capital investment since the need for thermal treatment chambers is eliminated.

The high dielectric characteristics of polymer concretes (see Chapter 5) make for high efficiency in using high-frequency (HFC) and superhigh frequency (SHFC) currents for accelerated hardening of small numbers of polymer concrete products. The heated material is described essentially by two parameters: dielectric constant ϵ and tangent of dielectric phase angle tg δ. The electrical energy liberated in the form of heat is proportional to the product of these two parameters and is called the factor or coefficient of loss K:

$$K = \epsilon \ \text{tg} \ \delta. \qquad \ldots (9.1)$$

The specific energy liberated per cubic centimetre of the material as a result of dielectric losses is determined by the following equation:

$$P = 5.56 E f \epsilon \ \text{tg} \ \delta \times 10^{-7}, \qquad \ldots (9.2)$$

where E is the voltage of the electric field in the material, kV/cm; and f the frequency, Hz.

It follows from equation (9.2) that if the material is homogeneous and the electrical field in it is uniform, the energy liberation and hence heating will proceed uniformly throughout the mass of the material. The energy liberated in the material is proportional to the square of the voltage and the frequency of the electrical field, i.e., depends on the field parameters. But, it is also proportional to the dielectric constant and tangent of dielectric phase angle of the material, i.e., depends on the electrical properties of the material.

It should be borne in mind that every material has a limiting value of field voltage which, if exceeded, may cause an electrical breakdown. In practice, to avoid such breakdowns, the working voltage is taken as 1.5 to 2 times less than the breakdown voltage.

Investigations showed that, when using industrial HFC generators, the duration of complete hardening of polymer concrete cubes with 50 mm sides is 25–30 min. A shortcoming of the method using HFC is the comparatively high power consumption and hence the use of HFC generators in the industry may be recommended mainly for hardening the control samples.

Hardening of polymer concrete specimens in SHFC units occurs in 3–4 min. This consumes comparatively less electrical energy compared to HFC heating.

The above methods of thermal treatment of polymer concrete products and structures suggest a method of greatly reducing the power consumption in this power-intensive operation. At the same time, it should be pointed out that, for thin-walled structures of small weight and large heat transfer surface, most of the above methods of thermal treatment (except SHFC heating) are not sufficiently effective. Such structures are: decorative finishing panels, window sill boards, staircases, decorative shapes, etc. Therefore, finding essentially new methods of economising power consumption in the stage of thermal treatment is an extremely urgent task.

One of the highly promising methods for resolving this problem is the utilisation of solar energy in the southern countries. Without going into detail about the vital possibilities and economic usefulness of utilising solar energy for thermal treatment of cement concretes, since this subject has been extensively reviewed in several works, it should be pointed out that, unlike cement concretes, polymer concretes require dry heating and solar energy in this respect is best suited.

It is known that the maximum amount of solar energy is received in regions from the equator to the 50° parallel. On sunny days, the amount of solar energy per sq. m is: at the equator over 5, at 35° latitude 4.2, and at 50° latitude 3.3 kW/h.

Studies in the field of utilising solar energy for thermal treatment of polymer concrete products are still lacking. However, the data given in [23] point to the high efficiency and the universal applicability of units utilising solar energy and the excellent physical and mechanical characteristics of the articles produced.

Fig. 9.22 shows the main layout of a solar chamber. In the course of a solar day, the temperature in such a chamber varies from 60° in the morning to 90°C in the day. This temperature range is entirely adequate for hardening thin-walled articles in the course of movement of the mould inside the chamber.

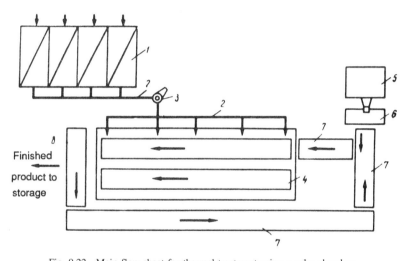

Fig. 9.22. Main flow sheet for thermal treatment using a solar chamber:

1—solar chamber (solar air heaters); 2—air ducts; 3—forced-draught fans; 4—polymerisation chamber; 5—concrete mixer; 6—vibration platform; 7—roller conveyors; 8—point of dismantling and cleaning moulds.

9.9. Quality Control of Polymer Concrete Products and Structures

The quality of polymer concrete products and structures depends directly not only on the quality of the raw materials but also, to no mean extent, on thoroughly conforming to the prescribed regimes and procedures of the technological process. Therefore, control of the quality of polymer concrete should provide for testing the quality of all the constituent components and the correctness of dosing and regimes of mixing, compaction and hardening.

The colour of the polymer concrete mix during mixing, moulding and heating in the process of hardening reflects the normal quality characteristics of the mix, while the increase in hardness and strength during the course of hardening and the strength characteristics and homogeneity after hardening represent its normal quality characteristics.

After adding benzenesulphonic acid and good mixing, a polymer concrete mix based on FAM (furfural-acetone) resin should be dark violet in colour. Yellowishness or emerald-coloured spots signal poor mixing. Polymer concrete mixes based on polyester, carbamide, phenol-formaldehyde and other resins after adding the appropriate amounts of hardeners and good mixing should be homogeneous in colouration and devoid of lumps.

After 15–20 min of retention in the mould, the polymer concrete mix begins to warm up. In structures of volume 0.2 m^3 or more, the self-heating temperature goes up to 70–90°C. Poor heating or absence of heating indicates a poor-quality resin or hardener, incorrect dosing, or high moisture content of fillers and aggregates.

For operational control of polymer concrete quality, primarily its strength characteristics, the usually adopted thermal treatment (18–24 h) is not suitable since the results obtained in this case cannot be used for correcting the technological process. For comparatively rapid thermal treatment, the use of HFC generators of capacity 2–4 kW or SHFC furnaces are recommended in large shops or factories.

For hardening polymer concretes by high frequency current, the heating of the material should proceed uniformly all through the body; the duration of total hardening of polymer concrete cubes with sides of 50 mm should be 25–30 min when using HFC or 3–4 min when using SHFC. After cooling, the cubes should be tested, the results used to assess the technological process and an appropriate correction made in the process if necessary.

To determine the strength grade and homogeneity of load-bearing structures, six cubes with sides of 100 mm and three prisms 70 mm × 70 mm × 210 mm are made simultaneously while casting the polymer concrete in the mould. The cubes undergo thermal treatment in the HFC generators or SHFC furnaces or by any other regime as adopted for commercial production.

The strength increase during hardening, modulus of elasticity, and also the homogeneity of polymer concrete (prisms 70 mm × 70 mm × 210 mm) are determined in an ultrasonic instrument.

9.10. Safety Requirements

When designing shops and factories for producing polymer concrete products and structures, appropriate measures should be implemented for ventilation, provision of scavenging equipment, waste utilisation and fire-fighting techniques, bearing in mind the toxicity, fire and explosion hazards, and also the maximum permissible concentrations (MPC) of these substances in the atmosphere.

Table 9.1 gives the basic characteristics of the deleterious constituents liberated during the production process of polymer concretes in respect of the most frequently used materials.

Table 9.1. Characteristics of deleterious constituents liberated in the process of producing polymer concrete

Substance	MPC, mg/m^3		Flash point, °C	Concentration range for vapour ignition, % by volume	Aggressiveness to metal
	in the working zone atmosphere of production buildings	in the atmosphere of inhabited points			
Polymer concrete based on FAM (FA) (furfural acetone) resin					
Furfural	10	0.05	61	1.8–3.4	Not aggressive
Monofurfurylidene acetone	0.1	0.001	80	—	-do-
Acetone	200	0.35	−20	2.55–12.8	-do-
Sulphuric acid	1	0.1	—	—	Aggressive
Benzene	5	0.8	10.7	1.5–8.0	Not aggressive
Polymer concrete based on PN (polyester) resin					
Styrene	5	0.003	34	1.1–6.1	-do-
Cobalt naphthenate	5	0.003	34	1.1–6.1	-do-
Hydroperoxide of isopropyl benzene	1	0.001	—	—	-do-
Polymer concrete based on UKS resin					
Formaldehyde	0.5	0.03	—	7–72	Not aggressive
Aniline hydrochloride	3	0.03	—	—	Aggressive
Polymer concrete based on FAED (furan-epoxide) resin					
Epichlorohydrin	1	0.02	35	1.2–13.3	Not aggressive
Toluene	50	0.6	4	1.3–7.0	-do-
Furfural	10	0.05	61	1.8–3.4	-do-
Acetone	200	0.35	−20	2.55–12.8	-do-
Polyethylene polyamine	1	—	—	—	-do-
Polymer concrete based on MMA (methyl methacrylate) resin					
Methyl methacrylate	50	0.1	10	2.1–12.5	-do-
Polymer concretes of all types					
Dust containing up to 70% free SiO$_2$	2	0.15	—	—	-do-

To protect the air basin from contamination, the exhaust from the thermal treatment chamber should be utilised or burnt. The rest of the ventilation discharges should be taken to a height adequate for their dispersal such that their concentration in the atmosphere of the working zone in the production buildings is 0.3 of MPC and the level in the atmosphere of human habitation conforms to the maximum permissible concentration.

When producing polymer concrete products and structures, all the rules provided in the corresponding specifications for structural safety as also the hygienic rules for organisations carrying out technological processes should be observed.

In the production buildings, electrical equipment and electrical lighting should be earthed. During production, lighting a fire, use of blowpipes, welding and other operations that are likely to give rise to sparks and flames are prohibited.

Work should be carried out only after switching on the suction and exhaust ventilation. Should the ventilation system fail, work must be terminated at once.

In the thermal treatment chambers, after loading the polymer concrete articles or structures, the exhaust ventilation system should be operated continuously.

There should be systematic control of the atmospheric conditions in all plant buildings. Before permitting a worker to function independently, he should undergo a course of lectures and instructions on safety and fire-fighting techniques.

Workers should be provided with special garb and individual protective devices: overalls made from a thick fabric, rubberised hoods, rubber boots, gloves and gas masks for any emergency.

The shop should be provided with separate closets for storing clean clothes, linen, special garb and medicine chests, as well as wash basins and hot water showers. Operations involving the production and moulding of polymer concrete mixes should only be carried out while wearing rubber gloves.

10

Polymer Sulphur Concretes: New Efficient Materials

The application of new materials and technological processes capable of reducing labour and material content and improvement in the quality and durability of civil structures and products are the two important and promising directions in the contemporary building industry.

Damages to structures and technological equipment cause frequent repairs and equipment hold-ups and, correspondingly, involve huge losses of material resources. Building structures functioning in the permafrost zone and regions with high soil salinity, for example, in the region of Mangyshlak and Kara-Bogaz Gola Bay in the Fore-Caspian lowland, are exposed to considerable damage. To enhance the durability of buildings and installations working under such conditions, polymer concrete structures which possess high strength, density and fairly universal chemical stability are recommended [62].

However, the comparatively high cost of the polymer materials used as binders in producing polymer concretes restrains their use in many cases. Therefore, scientists have been paying more attention to sulphur as an effective binder.

The use of sulphur in construction either as a paste or plaster was known even in the 19th century. In 1859, A.H. Right was granted a patent in which the use of sulphur plaster was indicated for fixing foundation bolts [58]. Sulphur pastes and plasters were even used in Russia in the last century for sealing the joints in stone masonry and were found to be much more effective for embedding metallic railing supports of staircases and metal joints of stone structures than molten lead.

Later, with the advancement of anticorrosion services, sulphur pastes and plasters, called sulphur cements, found application for sealing the joints in the refractory linings of various containers, apparatus in the chemical industry and building structures with acid-resistant plasters used under conditions of exposure to various aggressive media. Examples of sulphur cement compositions are shown in Table 10.1.

The ultimate compressive strength of sulphur cements is not less than

Table 10.1. Average compositions of sulphur cements

Constituent	Composition, % by weight		
	1	2	3
Technical sulphur	58	60	67
Mineral fillers	40	36	—
Coke	—	—	31
Plasticiser (Thiocol or thermoprene)	1–1.2	1–4	1–1.2

30 MPa and the tensile strength 2–2.5 MPa. Sulphur cements are stable in most mineral acids (except for fluoric and nitric acids) and solutions of mineral salts but are not stable in alkalis and many organic solvents (carbon disulphide, etc.).

In spite of the comparatively high effectiveness of sulphur cements in anticorrosion treatment practice, their total demand is insignificant compared to other materials; yet the demand for chemically stable and cold-resistant materials has been rising steadily. Therefore, many scientists are studying sulphur once again in order to widen the scope of its application in the various branches of the building industry.

Let it be also noted that, in recent decades, the production of natural and secondary sulphur obtained during oil and natural gas refining has been rising steadily. According to the data for 1985, the world production of sulphur was 53.75 million tonnes. This includes (in million tonnes) 14.5 in the USA, 6.5 in Canada, 5.4 in Poland, 2.9 in Japan and 5.77 in the Soviet Union [64]. Canada, Poland, and the USA produce more sulphur than is consumed within the respective countries.

Sulphur represents a typical thermoplastic material that is highly compatible with various modifiers and can transform into a polymer state under certain conditions. In size composition of fillers and aggregates, sulphur concretes approximate polymer concretes. The basic characteristics of the former are determined by the properties of the sulphur binder, type of mineral constituents, correct proportion of the different fractions, and type and amount of modifiers. Thus, it is possible to fix the basic data for sulphur concretes.

In world practice, three main directions for the use of sulphur in the building industry can be identified: production of sulphur-asphalt concretes for road construction; production of sulphur concretes for various purposes; and impregnation of cement concretes with molten sulphur.

Only sulphur concretes which approximate polymer concretes are examined in this chapter.

10.1. Basic Properties of Sulphur

It is known that sulphur has an unusual molecular structure since it is characterised by a large diversity of polymorphic modifications. So far, some thirty allotropes of sulphur have been isolated but most of them have not been adequately studied and no unified classification of them is available. But of the diverse modifications of sulphur, four are of primary interest to researchers: rhombic $S_\alpha(S_\delta)$, prismatic $S_\rho(S_\delta)$, cyclic ring $S_\lambda(S)$ and polymer $S_\mu(S_\infty)$. All of these allotropes have been well studied and their physical and chemical properties reported [16, 49, 67, 77]. The technical properties of sulphur, according to the data of various researchers, do not always coincide and are even contradictory in some cases. This is explained by the difficulty of producing specimens of stable dimensions due to the high shrinkage of sulphur, varying content of H_2S in sulphur and other factors. Table 10.2 gives the main physical properties of sulphur required when formulating sulphur concrete compositions and its average technical characteristics.

Table 10.2. Physical and technical characteristics of technical sulphur

Index	Temperature, °C		
	20	122	150
Density, g/cm³	2.1	1.96–1.99	1.6–1.81
Strength, MPa:			
compressive	18–20	—	—
bending	5–7	—	—
Modulus of elasticity, MPa $\times 10^3$	13–14	—	—
Microhardness, MPa	710–720	—	—
Coefficient of temperature deformations, $10^{-6}/°C$	5.7	11.8	163
Thermal conductivity, W/(m-°C)	0.27	0.13	0.14
Viscosity, Pa-sec $\times 10^{-3}$	—	11–12	6.5–7
Surface tension, N/cm $\times 10^{-5}$	—	—	0.055

Depending on the amount of H_2S contained in sulphur, ambient temperature, and the duration elapsing after sample preparation, the strength characteristics of sulphur may vary widely. According to [69], at a high H_2S concentration in the range of 200–250 parts per million parts of sulphur, the strength of seven-day-old sulphur specimens decreases by 80–90% compared to the strength of the control specimens free of H_2S. The mechanism of the effect of H_2S on the strength characteristics of sulphur has not been adequately studied. After 40 days of storage of specimens containing H_2S, a recovery of their strength characteristics was noticed. The strength recovery of sulphur specimens is associated with the thermal instability of hydrogen sulphide which decomposes with time.

According to [69], the strength variation of sulphur relative to temperature and duration of storage after preparation is extremely complex. Initially, as the temperature rises from $-100°C$ to $40°C$ and with the increasing duration of storage to 15–20 days at normal temperature, the strength of sulphur specimens increases, but on further increase of temperature and duration of storage, strength diminishes (Fig. 10.1).

220

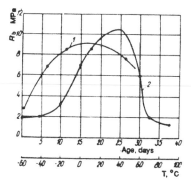

Fig. 10.1. Kinetics of strength variation of sulphur relative to age (1) and temperature (2) of specimens.

The investigations carried out at the Reinforced Concrete Research Institute and other organisations showed that the dependence of the strength of sulphur on temperature and duration of storage after preparation has a different pattern.

10.2. Polymer Sulphur Concretes

Sulphur concretes represent compositions consisting of mineral fillers and aggregates, sulphur binder and various modifying additives. In strength characteristics, sulphur concretes occupy an intermediate position between cement concretes V25 to V30 and high-strength polymer concretes. The technology of producing sulphur concretes differs very little from that of producing asphalt concretes.

According to the classification adopted in the USSR, sulphur concretes containing polymer-modifying additives are called *polymer concretes* on analogy with polymer cement concretes.

The theoretical principles of selecting the optimum compositions of polymer sulphur concretes should be based on the theory of rational structure formation of such systems and directed towards obtaining maximum density at least permissible consumption of sulphur binder while maintaining workability, high strength characteristics, and other physical and mechanical properties of the material.

Three basic structures can be identified among polymer sulphur concretes on analogy with polymer concretes: microstructure of sulphur paste, mesostructure of sulphur plaster and macrostructure of sulphur concrete.

It should be pointed out that in sulphur concretes as among polymer concretes, the specific surface of fillers constitutes 98–98.5% and of aggregates 1.5 to 2%. Therefore, the main structure-forming reaction among sulphur concretes arises in the stage of mixing the molten sulphur with finely dispersed fillers, i.e., in the stage of formation of the binding paste.

Researches carried out at the Reinforced Concrete Research Institute and L'vov Polytechnical Institute [80] showed that, depending on the extent of filling the molten sulphur with mineral flour, the ultimate compressive strength of specimens varies from 18–20 MPa without the filler to 50–60 MPa at 200–250%

of the filler. Further increase in filler leads to a sharp reduction in strength (Fig. 10.2.).

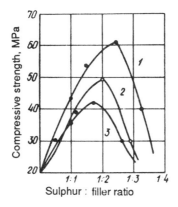

Fig. 10.2. Variation of compressive strength of sulphur pastes in relation to content of filler:

1—according to M.V. Paturoev's data; 2—according to E.V. Yur'eva's data; 3—according to V.A. Eremina's data for sulphur pastes with fillers of volcanic slag.

The distinctly manifest extremal character of strength variation in relation to the extent of filler confirms, in general, the similar trend obtained earlier for polymer pastes although the mechanism of their hardening differs. When producing sulphur pastes, more uniform crystals are formed on the surface of mineral fillers as the sulphur cools. These crystals are much smaller than in the sulphur body free of filler. At optimum filler content, almost the entire sulphur is transformed into a more uniform finely crystalline state. Further, a film of optimum thickness is formed in the sulphur paste around the filler grains and the integral adhesion bonds between the sulphur binder and the surface of the fillers and aggregates increased. The reduction of sulphur crystal size is strikingly confirmed in photographs taken under scanning microscope jSM-255. All of these factors are inalienably interrelated and ensure a very significant increase of strength at optimum filler content [80, 81].

Increasing the filler content beyond the optimum level leads to a significant increase in stiffness of the mix and inadequacy of sulphur binder for total contact with the filler surface and results in a subsequent strength reduction of the composition.

Thus, an extremely important principal was discovered which, with the help of theoretical assumptions and fundamental premises of the authors' theory of structure formation of polymer concretes, enabled development of the basic premises of a corresponding theory of structural formation of sulphur concretes by seeing them as a successive formation of different structures.

Microstructure of sulphur pastes

Among the main structure-forming factors that determine the properties of sulphur pastes are the type and dispersibility of fillers, degree of filling of the system, nature of adhesion bonds between the sulphur binder and the filler surface, and the extent of internal stresses arising in the system 'sulphur-filler'.

Research carried out at the Reinforced Concrete Research Institute and other organisations revealed that the optimum values of specific filler surface should lie in the range 3000 ± 500 cm^2/g and the optimum degree of filling the sulphur with mineral powder from 1:2 to 1:2.5. Investigations of the system 'sulphur-filler' (quartz and andesite flour, fly ash, etc.) showed that the extent of adhesion bonds is determined mainly by the physical bonds. It should further be remembered that sulphur by itself, without fillers, is capable of structural variations, including transition from one allotropic state to another (for example, sulphur S_7 at 37°C and sulphur S_6 at 50°C transform into sulphur S_8, etc.) under the influence of external factors (temperature drops, atmospheric effects, etc.). This transformation causes additional internal stresses and leads to a reduction of strength characteristics and in several cases to peeling and damage of the material entity.

The high adhesion bonds at optimum level of filling with mineral fillers greatly inhibit any adverse structural modifications of the sulphur binder but cannot totally exclude them. Therefore, to improve the structural stability of the sulphur binder and to improve the workability and durability of sulphur pastes and ultimately of sulphur concretes, various modifying agents are added to the compositions.

Among the most effective modifying additives are Thiocol, chloroparaffin, dicyclopentadiene, etc. at 1–5% of the weight of sulphur. These modifiers reduce the initial short-term strength by 12–15% but simultaneously stabilise the structural state of sulphur, plasticise the sulphur-concrete mix in the course of production and reduce the brittleness of the composition after hardening. Chloroparaffin ChP-1100 is not only a modifying additive but also a good fireproofing compound.

The next characteristic of a correctly prepared sulphur paste is the reduced content of air pores. In the course of producing the sulphur paste, modifier and mineral filler containing a significant amount of locked-in air are added to the molten sulphur. On thorough mixing of the filler with the molten sulphur at 145–150°C, degassing occurs and the bulk of air carried in by the filler is driven out.

Mesostructure of sulphur plasters

Unlike sulphur pastes, sulphur plaster additionally contains sand of grain size 0.15–3 mm. In this case, the decisive factors for optimum mesostructure are the maximum density of packing of the sand grains and the high adhesion bond of sulphur paste with sand. Experience has shown that sulphur paste possesses high adhesion to different types of sands and that the packing density in the above range of sand grains is about 66%, the corresponding voids being 34%. It is these voids that should be filled with an appropriate amount of sulphur paste.

Macrostructure of sulphur concretes

Sulphur concretes contain all the three fractions of fillers and aggregates— mineral flour, sand and rubble. On analogy with sulphur plasters, the factors determining the macrostructure of sulphur concretes are the same: maximum density of packing of aggregate grains and the high adhesion bond between the sulphur paste and the aggregate. Further, aggregates of varying size composition are used.

By using the resultant dependence of the structural strength of sulphur pastes on the extent of their filling with mineral fillers, data on the effect of modifiers on the technological and strength characteristics of sulphur compositions, theory of structure formation of polymer concretes, and the method of selecting their optimum compositions [80, 81], a method was developed for selecting the optimum compositions of heavy sulphur concretes. The principle of this method is as follows. Based on the type and specific surface of the mineral filler, the maximum permissible degree of filling the molten sulphur with the mineral filler is determined experimentally, i.e., the optimal composition of the sulphur paste is determined. Having determined the optimal degree of the filler and knowing the specific surface of the filler, the mean effective film thickness of the sulphur binder and the consumption of the binder for a given amount of the sulphur paste are determined on analogy with polymer concrete. Since the viscosity of sulphur at 145–150°C is almost constant and modifiers are added to the composition, the theoretical equations assume the following form:

$$\delta = m_b \alpha / S_{sp} m_f \rho_b \qquad \ldots (10.1)$$

and

$$M_p = \alpha (S_f m_f \rho_b \delta) 10^{-3}, \qquad \ldots (10.2)$$

where δ is the film thickness of the binder, cm; m_b the weight of the binder, g; S_{sp} the specific surface of the filler, cm^2/g; m_f the weight of the filler, g; ρ_b the density of the binder, g/m^3; α a coefficient allowing for the plasticisation of sulphur; and M_p the optimum weight of the paste, kg.

For known size composition of sand and aggregate, the optimal amount of the binder for heavy sulphur concrete is determined using the following equation:

$$M_{p.b} = [K(S_1 m_1 + S_2 m_2) \rho_b \delta] 10^{-3}, \qquad \ldots (10.3)$$

where $M_{p.b}$ is the optimum amount of the sulphur binder, kg; S_1 and S_2 the area of the specific surface of sand and aggregate, cm^2/g; m_1 and m_2 the weight of sand and aggregate, kg; and K the separation factor.

By using the above relationships, the optimal compositions of heavy sulphur concretes containing not more than 10–12% sulphur were selected first.

The test results (Table 10.3) showed that the scientifically selected compositions of heavy sulphur concretes have very high strength characteristics and such concretes can be satisfactorily used for making critical chemical-resistant load-bearing structures. These tests confirmed that the theory of structural formation of polymer concretes developed at the Reinforced Concrete Research Institute is quite universal and its main premises are wholly applicable for selecting the optimal compositions of sulphur concretes.

Table 10.3. Strength of sulphur concrete cubes under compression (specimen size 70 mm × 70 mm × 70 mm)

Composition number	Breaking load, kg	Compressive strength, MPa
1	39,500	80.6
2	38,600	78.9
3	38,225	78
4	37,975	77.5

It should also be pointed out that if the temperature of molten sulphur is raised to 180–200°C and cooled sharply thereafter, the allotropic modification of sulphur S_{∞} known as 'polymer' or 'insoluble sulphur' is formed. Such sulphur contains up to many tens of thousands of atoms in the polymer molecule.

An advantage of polymer sulphur over crystalline sulphur is its insolubility in known organic solvents and its ability to develop far less internal stresses on transition from liquid to solid state while its melt possesses much greater adhesion properties compared to the corresponding properties of crystalline sulphur melt. However, being a metastable allotrope, over the course of time it transforms at normal temperature into the crystalline modification with the formation of cyclic molecules.

To eliminate this phenomenon, stabilisers are added to the sulphur melt. These include red phosphorus together with iodine, dicyclopentadiene, etc., which stabilise sulphur in the polymer state.

The production of 100% polymer sulphur is a highly energy-consuming and laborious process. At the same time, investigations have shown that to obtain high-quality sulphur concretes, the polymer sulphur content in the sulphur binder can be in the range 50–60%. In this case, the amount of polymer sulphur in the melt is controlled by temperature and duration of heating and type and amount of modifiers.

Conforming to the adopted terminology, sulphur concretes produced by using polymer sulphur belong to the class of polymer concretes based on inorganic polymer and are called sulphur polymer concretes.

An economic evaluation of the application of polymer concretes based on sulphur binder containing 50–60% polymer sulphur showed that even at this level of polymer sulphur, the technological process is significantly complicated and the cost of sulphur polymer concrete goes up by roughly 30%. Therefore, in each actual case, the use of polymer sulphur for producing sulphur polymer concretes should be justified on economic considerations.

10.3. Physical and Mechanical Properties of Polymer Sulphur Concretes

Since the sulphur binder is an inorganic thermoplastic material, for correct evaluation of the test results using polymer sulphur concretes, on analogy with polymer concretes, external stress σ, deformation ϵ, temperature t, and test duration (time) t should be taken into consideration.

At normal stable temperature and constant loading rate, the short-term strength characteristics of polymer sulphur concretes is determined by the equations: $\epsilon = f(\sigma)$ and $\sigma = f(\epsilon)$. Further, tests are carried out at constant stress amplitude ($\sigma = const$) or at constant deformation amplitude ($\epsilon = const$). Equations $\epsilon = ft$ and $\sigma = ft$ are resolved by carrying out long-term tests.

Statistically processed results of testing sulphur and polymer sulphur concretes at normal temperature are given in Table 10.4.

Table 10.4. Results of determining strength characteristics

Type of test	Sulphur concrete without modifiers	Polymer sulphur concrete modified with chloroparaffin
Compressive strength, MPa (cubes, sides 70 mm and 100 mm)	75	70
Transverse tensile strength, MPa (prisms 40 mm × 40 mm × 160 mm)	10	12
Elongation on cleavage, MPa (cubes, sides 100 mm)	2.15	2.3
Modulus of elasticity under compression, 10^4 MPa (prisms 70 mm × 70 mm × 280 mm)	4.5–5.2	4.2–4.5

The effect of the plasticiser on the brittle strength of polymer sulphur concrete was evaluated by the method of V.I. Shevchenko, whereby an equilibrium breaking of the specimen is effected, the deformation diagram complete with the down-trending branch recorded and the critical length of the equilibrium fissure calculated, which represents an objective measure of the brittleness of the material [71].

It is known that the machines used for testing the specimens for strength are made to a rigid or close to a rigid loading regime which leads to brittle, i.e., sudden breaking of the material. As a result, when testing concrete specimens,

226

only a portion of the complete deformation diagram can be obtained, i.e., up to the ultimate strength of the material.

Under compatible conditions of the rigidity of the testing machine and the specimen dimensions and at constant rate of deformation or using special attachments to standard machines, equilibrium breaking of the specimen can be effected and the deformation diagram complete with the down-trending branch recorded.

A special device (Fig. 10.3) provides, under conditions of stable breaking of the specimens, more reliable results of the required parameters. The principle of this method is as follows. In the mid-portion of a specimen 40 mm × 40 mm × 160 mm, a transverse groove to a depth of 0.5–0.75 of the cross-section is cut by a diamond saw. Specimen 4 is set up on two supports and punch 2 is brought above it through cushion 3. The punch is rigidly fixed to precalibrated ring 6.

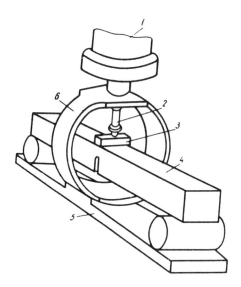

Fig. 10.3. Device for testing specimens under conditions of stable breaking:

1—loading element; 2—punch; 3—cushion; 4—specimen; 5—base; 6—power ring.

Force P transmitted in this device to the specimen and the deformation f corresponding to it are recorded by special gauges (not shown in the sketch) on a recording double-co-ordinate potentiometer. The area of the diagram within the co-ordinates 'force-bending' represents the work performed during breakage. This work calculated per unit surface of breakage gives the effective breaking energy G_{ic}. This value G_{ic} includes not only the thermodynamic surface energy of breaking but also dissipation (loss) of energy due to irreversible deformations on the fissure front during its advance.

The results of testing sulphur concretes of different compositions are shown in Fig. 10.4. Having determined first the value of modulus of elasticity E_d, K_f is calculated as

$$K_f = \sqrt{E_d G_f} \, . \qquad \qquad \dots (10.4)$$

The parameters of the complete deformation diagram, transverse tensile strength $R_{t.t}$, dynamic modulus of elasticity on elongation under bending E_d, and critical lengths of equilibrium fissure l_{fis} are shown in Table 10.5.

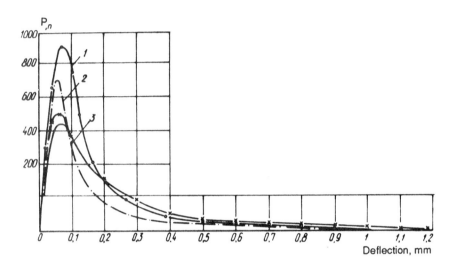

Fig. 10.4. Total equilibrium deformation of sulphur concrete:

1—sulphur concrete without modifier; 2—fine-grained sulphur concrete without modifier; 3—polymer sulphur concrete modified with chloroparaffin.

Fine-grained sulphur concrete (composition 2, see Table 10.5) has much lower G_f and K_f values compared to sulphur concrete with aggregate (composition 1). Thus, these tests confirm the data that the addition of aggregate to the composition of sulphur concretes results in their increased fracture resistance.

Based on the condition of balance of elastic energy accumulated in the sample and the breaking energy of the material, the following equation was given in [71]:

$$l_{fis} = E_d G_f / R_{t.t}^2 \, . \qquad \qquad \dots (10.5)$$

This equation characterises the critical length of the equilibrium fissure and represents a measure of the brittleness of the material; the lower the value of l_{fis}, the more brittle the material and vice versa.

Table 10.5.

No. of composition	Composition of concrete	$R_{t.t}$, MPa	$E_d \times 10^3$, MPa	G_f, J	K_f, MPa$^{3/2}$	l_{fis}, mm
1	Sulphur concrete with aggregate	13.87	45.1	232.4	3.24	54.5
2	Fine-grained (sandy)	11.48	42.8	152.2	2.55	49.4
3	Sulphur concrete with aggregate modified with ChP-1100	10.44	37.52	208.8	2.8	71.6

The critical length of the equilibrium fissure calculated using equation (10.5) for fine-grained concrete is 49.4 mm and for sulphur concrete with aggregate 54.5 mm, and for concrete modified with chloroparaffin 71.6 mm. Thus, modified sulphur concrete has very low brittleness compared to fine-grained sulphur concrete (by 45%) and sulphur concrete containing aggregate (by 31.4%).

The coefficient of temperature deformations (CTD) is one of the important criteria for evaluating the material deformability. The CTD value is of particularly great importance when evaluating the unified performance of sulphur concrete with the reinforcement in reinforced structures exposed to high temperature drops and also when using sulphur concrete as a protective coating or in two or three layers when some types of materials function in a unified manner.

The CTD of sulphur varies in a very wide range depending on temperature and bears an extremely complex character (Fig. 10.5). However, in the temperature range 15–100°C, this dependence is rectilinear and the CTD of sulphur varies from 5.7×10^{-6} at 15°C to 11.7×10^{-6} at 100°C; the CTD of sulphur changes sharply only beyond 100°C. Data available in the literature on the CTD of sulphur at negative temperatures are very scant and contradictory.

Since the thermal stability of sulphur concretes falls in the region of 80°C, reliable data on the CTD of sulphur concretes are necessary only in this range of positive temperatures.

The CTD of sulphur concretes was determined by the usual method based on the assumption of a direct dependence of the variation of deformation on temperature in a given temperature range using the following equation:

$$\alpha = (1/l)(\Delta l/\Delta T) , \qquad \qquad \dots (10.6)$$

where α is the CTD, 1/deg; l the initial length of the specimen, mm; Δl the change of specimen length during the test, mm; and ΔT the temperature variation of the material, °C.

The mean arithmetical value of the CTD of sulphur concrete without modifier was 10.45×10^{-6} 1/deg and of sulphur concrete with modifier ChP-1100 10.3×10^{-6} 1/deg.

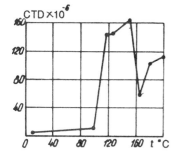

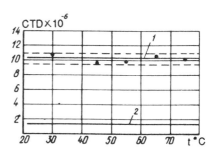

Fig. 10.5. Variation of the coefficient of temperature deformations (*CTD*) of sulphur relative to temperature in the range 15–200°C.

Fig. 10.6. Coefficient of temperature deformations (*CTD*) in the temperature range 15–80°C:

1—sulphur concrete; 2—invar (alloy of nickel and iron).

Thus, the *CTD* of sulphur concretes with and without modifiers is practically similar and falls in the zone of the confidence range of experimental accuracy (Fig. 10.6).

Investigations carried out on determining the *CTD* showed possible compatibility and good unified performance of sulphur concrete with steel reinforcement in load-bearing structures and with cement concrete given protective coatings of sulphur concrete or using multiple layers of it in structures.

10.4. Effect of Temperature on Properties of Polymer Sulphur Concretes

Along with many positive features, polymer sulphur concretes suffer from several vital drawbacks, of which the primary ones are the comparatively low thermal stability and inflammability (combustibility). These important characteristics have not been studied adequately and the information provided by several publications on this subject is contradictory.

Since sulphur is a typical inorganic thermoplastic material, it may be assumed that the effect of low temperatures on polymer sulphur concretes should be the same as on thermoplastic polymers.

The strength characteristics of polymer sulphur concretes were determined at low temperatures from 20 to −60°C at 20°C intervals on prism specimens 40 mm × 40 mm × 160 mm and cubes with sides of 70 mm. The samples were cooled in a thermal vacuum chamber. The time required for equalising a given temperature throughout the section of the specimen was determined using thermocouples fixed at the centre of the control specimen placed in the thermal vacuum chamber together with the specimens under test.

230

The test results confirmed the assumptions and showed that as the temperature decreases from 20 to −60°C, the strength characteristics of polymer sulphur concrete increase from 52 to 68 MPa (Fig. 10.7). Experimental data confirmed the possibility of using various building structures made of polymer sulphur concretes; such structures could be put to successful use in the regions of the Far North.

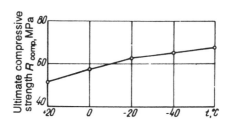

Fig. 10.7. Dependence of variation of ultimate compressive strength R_{comp} of polymer sulphur concrete on temperature reduction.

It is known that low thermal stability restricts the zone of application of load-bearing structures made of polymer sulphur concretes in industrial establishments in which high temperatures prevail. However, the maximum permissible positive temperature at which the structure does not lose more than 20% of its bearing capacity has not been determined so far.

Investigations on determining the strength deformation characteristics of different types of polymer concretes based on thermosetting oligomers showed that all of them conform to the common pattern, i.e., as the temperature rises from 20 to 100°C, the ultimate strength and modulus of elasticity fall in direct proportion. On cooling to 20°C, there is a near-total restoration of these characteristics.

An analysis of the graphs showing the variations of the ultimate strength of polymer sulphur concretes in compression and transverse tensile strength tests showed, for the first time, an extremely interesting anomalous property that is typical of sulphur concretes exclusively [80]. This is an increase in strength on raising the temperature from 20°C to 75–80°C and its perceptible decrease only with a further increment in temperature (Fig. 10.8).

The variation of modulus of elasticity relative to temperature variation in the range 20–100°C was determined on prism specimens 100 mm × 100 mm × 400 mm in a special unit consisting of a box-type supply line for stage-wise heating of the specimens with a supporting table and extenders for measuring deformations, automatic potentiometer to which Chromel-Copel thermocouples were connected for recording the temperature in the muffle and the specimen, and devices for controlling temperature and the press.

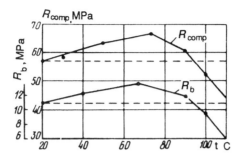

Fig. 10.8. Dependence of variation of ultimate compressive and transverse tensile strengths of polymer sulphur concretes on temperature increment.

Fig. 10.9 gives the average values of the modulus of elasticity under compression relative to temperature increment. On raising the temperature from 20 to 75°C, the modulus of elasticity, like the ultimate strength, rose from 50×10^3 to 62×10^3 MPa and a reduction of modulus of elasticity was noticed only beyond 75°C.

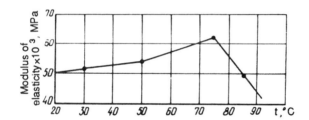

Fig. 10.9. Variation of modulus of elasticity relative to the increasing temperature.

The increase in strength and deformation characteristics of polymer concretes in the above temperature range exceeds the normal range of behaviour of such materials when testing under high temperature conditions. Therefore, it was necessary to thoroughly analyse the properties discovered and to seek an explanation, albeit only as a first approximation.

While studying the *CTD* of sulphur in a highly sensitive dilatometer, J. Miller (USA) noticed several deviations from proportionality in the *CTD* graph (Fig. 10.10, curve 1). These were more apparent on cooling the sulphur specimens to temperatures of liquid nitrogen (Fig. 10.10, curve 2). Further, smooth and comparatively insignificant volume changes were noticed at temperatures of 35 and 100°C and extremely high and sharp bumps at temperatures of 77 and 119°C. The volume change at 119°C was associated with the fusion of the prismatic form of the monoclinic modification of sulphur. J. Miller explained the change of volume at 100°C as a process associated with the fusion temperature of the rhombic form of sulphur and the temperature of the normal level of its existence which falls below 95.5°C. He explained the change of volume at 35°C as due to the transition from a glassy to a monoclinic form of sulphur.

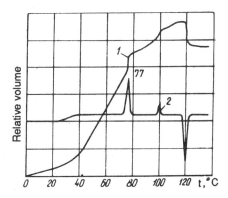

Fig. 10.10. Results of dilatometric measurements of sulphur relative to temperature.

The most important and interesting phenomenon, from our viewpoint, is the sharp change of volume at 77°C which, according to J. Miller, is the result of the reorganisation of the elastic sulphur into a new modification.

In our view, J. Miller's explanation of the processes occurring in sulphur at 35 and 77°C is not sufficiently convincing and does not reflect the actual importance of the phenomenon. The detected phenomenon of the anomalous change of strength and deformation properties of polymer sulphur concretes, with a certain amount of approximation, may be explained by the subsequent structural changes of the sulphur binder which occur under the influence of high temperatures. As already pointed out, sulphur represents a mixture of its various allotropic forms. At 20°C, apart from $S_\alpha (S_8)$, its composition may contain some forms of allotropic modifications including S_6, S_7, etc.

Allotropic modification S_7 is stable up to 39°C and, at this temperature, transforms into rhombic sulphur $S_\alpha (S_8)$. At 50°C, allotropic modification S_6 is also totally transformed into rhombic S_α sulphur.

Thus, on heating sulphur concretes roughly up to 80°C, the structural state is regulated and almost all of the sulphur is transformed into a uniform and more stable structure consisting of rhombic sulphur S_α. Further, on raising the temperature to 70–80°C, the defective crystals are restored in rhombic S_α and incomplete relaxation processes proceed intensely and reduce the internal stresses. These structural changes represent the main reasons for the significant increase in strength of sulphur concretes in the temperature range 20–80°C.

Some discrepancy in the temperature ranges with the data for pure sulphur could, in all probability, be explained by the high adhesion bonds of sulphur

with the surface of mineral fillers which extend and, to a certain extent, shift the temperature ranges of the structural reorganisation of sulphur binder.

Above 75–80°C, the effect of temperature softening and gradual recrystallisation of rhombic sulphur S_α into prismatic sulphur S_β commences. The latter possesses low density and strength compared to rhombic S_α and at 122°C sulphur fuses completely.

Thus, in Figs. 10.9 and 10.10, two distinct sections are noticed. In the first section, in the temperature range 20 to 80°C, an increase of strength and modulus of elasticity of sulphur concretes occurs and, in the second in the range of 80°C and above, intense reduction of strength characteristics and modulus of elasticity is noticed. It should further be pointed out that, at 90°C, the strength of sulphur concrete is roughly equal to the initial strength at 20°C.

These experimental data suggest with adequate justification that the maximum permissible operational temperature of load-bearing structures of sulphur concretes is equal to 80°C. However, unlike polymer concretes, at this temperature, there is not only no reduction in strength, but rather it actually increases by 20–25%.

10.5. Inflammability of Polymer Sulphur Concretes

As already pointed out, polymer sulphur concretes contain 10–13% sulphur with different modifiers while the rest (87–90%) comprises incombustible mineral fillers and aggregates. Thus, the very composition structure of polymer sulphur concretes ensures the minimum possible combustibility and inflammability.

Since polymer concretes in their mineralogical composition of aggregates and fillers and the amount of binder are most proximate to the compositions of polymer sulphur concretes, the method of determining the inflammability of polymer concretes can be rationally adopted for determining the inflammability of polymer sulphur concretes.

The results of testing sulphur concretes for inflammability in a ceramic tube (average value of three specimens) are given in Table 10.6. Sulphur concretes containing 13% sulphur underwent a weight loss of 10.36% and thus fell in the group of inflammable materials. On adding 2% chloroparaffin to the sulphur binder, the weight loss was 7% and such concretes fell in the group of not readily inflammable materials.

It is extremely interesting to note that on reducing the sulphur binder to 11% of the total weight of sulphur concrete, the weight loss after the test in a ceramic tube was 2.65% and thus even without adding fire-retarding agents, these sulphur concretes fall in the group of not readily inflammable materials, while the addition of 2% chloroparaffin reduced the weight loss of concrete to 0.76%.

Table 10.6. Results of determining the inflammability of sulphur concretes in a ceramic tube

Type of sulphur concrete	Sulphur content, %	Weight of specimen, g		Weight loss	
		before test	after test	g	%
Without modifier	13	671.96	598.96	73	10.36
With modifier (chloroparaffin)	13	654.43	608.13	46.53	7.01
Without modifier	11	633.2	626.15	17.05	2.65
With modifier (chloroparaffin)	11	656.65	651.6	5.02	0.76

Tests on flame propagation along the sample surface were carried out conforming to standard ISO 5658. The principle of the method lies in determining the flame propagation characteristics along the sample surface which is subjected to a single exposure of thermal radiation from a 13 kW radiation panel; the specimen is set at 45° to the gas burner flame.

Standard Distribution of Temperature from a Radiation Panel along the Specimen Model

Distance from end of specimen, mm	25	125	250	375	500	625
Temperature, °C	400	330	260	220	140	120

During the tests, the moment of burning of the specimen, the time in which the flame front traversed the transverse grooves made on the specimen at 25 mm intervals, cessation of flame propagation, maximum distance (in cm) traversed by the flame front, and the time of extinction of the flame were recorded.

The linear values of flame propagation (in cm) and the average rate of flame propagation (in cm/sec) over a length of 20 cm from the 'hot' edge of the specimen were taken as criteria for evaluating the flame propagation characteristics of the materials. The linear flame propagation is measured from the marks left by the flame on the specimen as a mean arithmetical value obtained in tests on three specimens with the result rounded to 1 cm. The average velocity of flame propagation is calculated as

$$V = 20/(t_{20} - t_i) , \qquad \ldots (10.7)$$

where t_i is the time before the inflammation of the specimen and t_{20} the time taken by the flame front to advance to the 20 cm mark.

The average arithmetical results of three tests were rounded to 0.1 cm/sec.

with the surface of mineral fillers which extend and, to a certain extent, shift the temperature ranges of the structural reorganisation of sulphur binder.

Above 75–80°C, the effect of temperature softening and gradual recrystallisation of rhombic sulphur S_α into prismatic sulphur S_β commences. The latter possesses low density and strength compared to rhombic S_α and at 122°C sulphur fuses completely.

Thus, in Figs. 10.9 and 10.10, two distinct sections are noticed. In the first section, in the temperature range 20 to 80°C, an increase of strength and modulus of elasticity of sulphur concretes occurs and, in the second in the range of 80°C and above, intense reduction of strength characteristics and modulus of elasticity is noticed. It should further be pointed out that, at 90°C, the strength of sulphur concrete is roughly equal to the initial strength at 20°C.

These experimental data suggest with adequate justification that the maximum permissible operational temperature of load-bearing structures of sulphur concretes is equal to 80°C. However, unlike polymer concretes, at this temperature, there is not only no reduction in strength, but rather it actually increases by 20–25%.

10.5. Inflammability of Polymer Sulphur Concretes

As already pointed out, polymer sulphur concretes contain 10–13% sulphur with different modifiers while the rest (87–90%) comprises incombustible mineral fillers and aggregates. Thus, the very composition structure of polymer sulphur concretes ensures the minimum possible combustibility and inflammability.

Since polymer concretes in their mineralogical composition of aggregates and fillers and the amount of binder are most proximate to the compositions of polymer sulphur concretes, the method of determining the inflammability of polymer concretes can be rationally adopted for determining the inflammability of polymer sulphur concretes.

The results of testing sulphur concretes for inflammability in a ceramic tube (average value of three specimens) are given in Table 10.6. Sulphur concretes containing 13% sulphur underwent a weight loss of 10.36% and thus fell in the group of inflammable materials. On adding 2% chloroparaffin to the sulphur binder, the weight loss was 7% and such concretes fell in the group of not readily inflammable materials.

It is extremely interesting to note that on reducing the sulphur binder to 11% of the total weight of sulphur concrete, the weight loss after the test in a ceramic tube was 2.65% and thus even without adding fire-retarding agents, these sulphur concretes fall in the group of not readily inflammable materials, while the addition of 2% chloroparaffin reduced the weight loss of concrete to 0.76%.

Table 10.6. Results of determining the inflammability of sulphur concretes in a ceramic tube

Type of sulphur concrete	Sulphur content, %	Weight of specimen, g		Weight loss	
		before test	after test	g	%
Without modifier	13	671.96	598.96	73	10.36
With modifier (chloroparaffin)	13	654.43	608.13	46.53	7.01
Without modifier	11	633.2	626.15	17.05	2.65
With modifier (chloroparaffin)	11	656.65	651.6	5.02	0.76

Tests on flame propagation along the sample surface were carried out conforming to standard ISO 5658. The principle of the method lies in determining the flame propagation characteristics along the sample surface which is subjected to a single exposure of thermal radiation from a 13 kW radiation panel; the specimen is set at 45° to the gas burner flame.

Standard Distribution of Temperature from a Radiation Panel along the Specimen Model

Distance from end of specimen, mm	25	125	250	375	500	625
Temperature, °C	400	330	260	220	140	120

During the tests, the moment of burning of the specimen, the time in which the flame front traversed the transverse grooves made on the specimen at 25 mm intervals, cessation of flame propagation, maximum distance (in cm) traversed by the flame front, and the time of extinction of the flame were recorded.

The linear values of flame propagation (in cm) and the average rate of flame propagation (in cm/sec) over a length of 20 cm from the 'hot' edge of the specimen were taken as criteria for evaluating the flame propagation characteristics of the materials. The linear flame propagation is measured from the marks left by the flame on the specimen as a mean arithmetical value obtained in tests on three specimens with the result rounded to 1 cm. The average velocity of flame propagation is calculated as

$$V = 20/(t_{20} - t_i) , \qquad \ldots (10.7)$$

where t_i is the time before the inflammation of the specimen and t_{20} the time taken by the flame front to advance to the 20 cm mark.

The average arithmetical results of three tests were rounded to 0.1 cm/sec.

with the surface of mineral fillers which extend and, to a certain extent, shift the temperature ranges of the structural reorganisation of sulphur binder.

Above 75–80°C, the effect of temperature softening and gradual recrystallisation of rhombic sulphur S_α into prismatic sulphur S_β commences. The latter possesses low density and strength compared to rhombic S_α and at 122°C sulphur fuses completely.

Thus, in Figs. 10.9 and 10.10, two distinct sections are noticed. In the first section, in the temperature range 20 to 80°C, an increase of strength and modulus of elasticity of sulphur concretes occurs and, in the second in the range of 80°C and above, intense reduction of strength characteristics and modulus of elasticity is noticed. It should further be pointed out that, at 90°C, the strength of sulphur concrete is roughly equal to the initial strength at 20°C.

These experimental data suggest with adequate justification that the maximum permissible operational temperature of load-bearing structures of sulphur concretes is equal to 80°C. However, unlike polymer concretes, at this temperature, there is not only no reduction in strength, but rather it actually increases by 20–25%.

10.5. Inflammability of Polymer Sulphur Concretes

As already pointed out, polymer sulphur concretes contain 10–13% sulphur with different modifiers while the rest (87–90%) comprises incombustible mineral fillers and aggregates. Thus, the very composition structure of polymer sulphur concretes ensures the minimum possible combustibility and inflammability.

Since polymer concretes in their mineralogical composition of aggregates and fillers and the amount of binder are most proximate to the compositions of polymer sulphur concretes, the method of determining the inflammability of polymer concretes can be rationally adopted for determining the inflammability of polymer sulphur concretes.

The results of testing sulphur concretes for inflammability in a ceramic tube (average value of three specimens) are given in Table 10.6. Sulphur concretes containing 13% sulphur underwent a weight loss of 10.36% and thus fell in the group of inflammable materials. On adding 2% chloroparaffin to the sulphur binder, the weight loss was 7% and such concretes fell in the group of not readily inflammable materials.

It is extremely interesting to note that on reducing the sulphur binder to 11% of the total weight of sulphur concrete, the weight loss after the test in a ceramic tube was 2.65% and thus even without adding fire-retarding agents, these sulphur concretes fall in the group of not readily inflammable materials, while the addition of 2% chloroparaffin reduced the weight loss of concrete to 0.76%.

Table 10.6. Results of determining the inflammability of sulphur concretes in a ceramic tube

Type of sulphur concrete	Sulphur content, %	Weight of specimen, g		Weight loss	
		before test	after test	g	%
Without modifier	13	671.96	598.96	73	10.36
With modifier (chloroparaffin)	13	654.43	608.13	46.53	7.01
Without modifier	11	633.2	626.15	17.05	2.65
With modifier (chloroparaffin)	11	656.65	651.6	5.02	0.76

Tests on flame propagation along the sample surface were carried out conforming to standard ISO 5658. The principle of the method lies in determining the flame propagation characteristics along the sample surface which is subjected to a single exposure of thermal radiation from a 13 kW radiation panel; the specimen is set at 45° to the gas burner flame.

Standard Distribution of Temperature from a Radiation Panel along the Specimen Model

Distance from end of specimen, mm	25	125	250	375	500	625
Temperature, °C	400	330	260	220	140	120

During the tests, the moment of burning of the specimen, the time in which the flame front traversed the transverse grooves made on the specimen at 25 mm intervals, cessation of flame propagation, maximum distance (in cm) traversed by the flame front, and the time of extinction of the flame were recorded.

The linear values of flame propagation (in cm) and the average rate of flame propagation (in cm/sec) over a length of 20 cm from the 'hot' edge of the specimen were taken as criteria for evaluating the flame propagation characteristics of the materials. The linear flame propagation is measured from the marks left by the flame on the specimen as a mean arithmetical value obtained in tests on three specimens with the result rounded to 1 cm. The average velocity of flame propagation is calculated as

$$V = 20/(t_{20} - t_i) , \qquad \qquad \ldots (10.7)$$

where t_i is the time before the inflammation of the specimen and t_{20} the time taken by the flame front to advance to the 20 cm mark.

The average arithmetical results of three tests were rounded to 0.1 cm/sec.

Based on flame propagation property, the structural materials can be subdivided into three groups: not propagating flame; moderately propagating flame; and intensely propagating flame.

Materials for which the linear flame propagation is less than 20 cm and the rate of flame propagation is equal to zero fall in the group of materials which do not support flame propagation. The group of moderately flame propagating materials covers materials for which these values do not exceed 57 cm and 0.5 cm/sec. Even if one of these values exceeds the above limit, the material is placed in the group of intensely flame propagating materials.

The results of tests (as mean values) showed that the inflammation of the specimens proceeded after 4.5 min and the flame front advanced by 20 cm in 10 min.

By substituting in equation (10.7) the experimental results, we derive

$$V = 20/(t_{20} - t_i) = 20/(600 - 270) = 0.06 \text{ cm/sec} .$$

Thus, the linear flame propagation is 20 cm while the flame propagation velocity is 0.06 cm/sec, i.e., although polymer sulphur concretes fall in the second group on the basis of these values, they essentially fall on the borderline between the materials pertaining to the first group which do not support flame propagation and the second group of materials which moderately propagate the flame.

The main physical and mechanical characteristics of sulphur concretes are given in Table 10.7.

Table 10.7. Main physical and mechanical properties of sulphur concretes

Index	Without modifier	With modifier ChP1100
Density, kg/m^3	2400	2400
Short-term strength, MPa:		
compressive strength	70–80	65–75
transverse tensile strength	12–13	10–11
tensile strength on cleavage	2.3–2.5	2.1–2.3
Modulus of elasticity under compression, MPa $\times 10^4$	4.5–5.2	4.2–4.5
Dynamic modulus of elasticity, MPa $\times 10^4$	5.12	4.47
Coefficient of temperature deformation, 1/°C	$(8-12) \times 10^{-6}$	$(8-12) \times 10^{-6}$
Water absorption after 24 h, %	0.06–0.07	0.05–0.065
Cold resistance, cycles	Over 300	Over 300
Thermal stability, °C	80–85	80–85
Abradability, g/cm^2	0.26–0.27	0.27–0.28
Tangent of dielectric phase angle		
(in dry condition), 10^{-2}	1.17	1.15
Dielectric constant (in dry condition)	4.24	4.5
Water impermeability	Impermeable	Impermeable

10.6. Prospects for Developing and using Polymer Sulphur Concretes in Construction

Apart from superior physical and mechanical characteristics, sulphur concretes possess many other favourable properties. These are: rapid gain of strength which is determined by the duration of setting of the sulphur concrete mix after its casting in the mould; possibility of reinforcing sulphur concrete structures with steel as well as glass-plastic reinforcement or minced fibreglass; unlike cement concretes, sulphur concretes can be reinforced with ordinary aluminoborosilicate fibreglass; and the possibility of re-using sulphur concretes by crushing the old or rejected structures and remoulding them.

The high strength characteristics, low water absorption, chemical resistance to acidic solutions, mineral salts and other aggressive media, besides the favourable features mentioned in the previous paragraph and also the comparatively low cost enable a wide range of sulphur concrete products and structures and provide rational fields for their application.

Among the popular types of sulphur concrete products and structures are supports or piers for pipes when laying oil and gas pipe lines across rivers and marshes. Such supports are used in Canada and other countries. For example, the Canadian firm Sulcon Concrete Ltd. produces precast supports of outer diameter 1830 mm, length 1850 mm, and weight 6250 kg which are used for laying pipe lines across rivers. Supports for pipes for laying across marshes are made in lengths of 1725 mm and weight 5150 kg with a different structural design to facilitate laying in a marshy locality.

The Chevron Riserg Co. and Chevron Chemical developed a method for facing earthen walls of irrigation canals using sulphur concretes. The coating is applied by guniting in layers of thickness 5–6 mm. Depending on the type and degree of soil compaction, the total thickness of the protective coating can go up to 20–40 mm.

For applying such protective facings, special mobile equipment capable of working under field conditions has been designed. Compared to the concrete facings, gunited protective coatings of fine-grained sulphur concretes reduce water in filtration and greatly cut short the construction duration.

Sulphur concretes have not been studied adequately yet nor have their potential possibilities been fully identified but the above brief review indicates that the possible uses of these new and progressive materials are manifold.

Mention must be made of the large amounts of sulphur-bearing wastes generated in many branches of the chemical industry, mineral fertiliser industries, non-ferrous metallurgy, etc. containing 15–20 to 60–70% technical sulphur.

The investigations of A.E. Nikitin (Reinforced Concrete Research Institute) showed that in spite of the diversity of such materials, they can be tentatively divided into three groups based on the form and content of mineral constituents: sulphur-bearing wastes with all the main fractions of fillers and aggregates (rubble, sand, mineral flour, etc.); sulphur-bearing wastes containing sands and mineral flour; and sulphur-bearing wastes with only finely dispersed fractions of mineral flour.

Sulphur-bearing wastes of the first group contain up to 30% technical sulphur and are typical of the chemical industry. Thus, works producing sulphuric acid alone generate 25,000 to 30,000 tonnes of such wastes annually.

Sulphur-bearing wastes of the second group are characteristic of the technical sulphur production industry by direct methods of its thermal recovery from native ores or by underground melting. In this case, the molten sulphur after separation enters settlers in which the mineral impurities are separated out. The sulphur content in such wastes goes up to 50%.

Sulphur-bearing wastes of the third group are characteristic of works which produce the purer grades of sulphur from technical sulphur. The sulphur content of these wastes goes up to 60–70%.

Typical compositions of the three types of sulphur-bearing wastes are shown in Fig. 10.11.

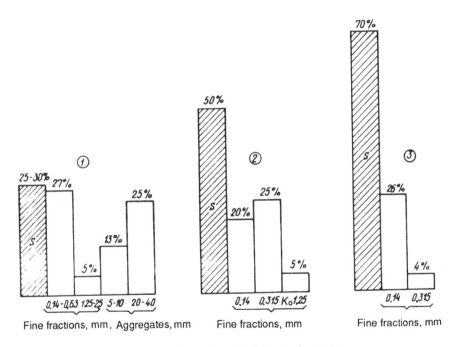

Fig. 10.11. Compositions of sulphur-bearing wastes.

The first group of sulphur-bearing wastes can be used to produce only less valuable products, for example, footpath slabs or tiles whose strength characteristics are 30–40% less compared to sulphur concretes based on technical sulphur.

However, even when producing such products, appropriate modifying agents must be added to the composition.

Since the size composition of mineral constituents in all the three groups of sulphur-bearing wastes is arbitrary and they contain a large amount of sulphur, especially the third group, the size composition of mineral aggregates should be corrected to bring it as proximate as possible to the optimum level for producing good quality products and structures utilising these waste products.

The technology of producing products and structures using sulphur-bearing wastes is very complex and the scatter of physical and mechanical characteristics of the finished products high. At the same time, it should be remembered that the utilisation of such wastes holds promise on grounds of their cheapness and easy availability.

Based on the complex investigations carried out, a plant has been designed to produce products and structures of polymer sulphur concretes with a capacity of 3000 m^3 a year and its construction has commenced at one of the non-ferrous metallurgical combines.

The technological flow sheet of this plant (Fig. 10.12) consists of six main sections: storage of raw materials; section for producing the sulphur binder with the modifying agent; drying, heating, and preparation of the fillers and aggregates; section for preparing and moulding the sulphur-concrete mix with preparation and heating of the mould, reinforcing frame, components for insertion and automated control panel; storage of finished products and section for processing the rejected articles; and section for gas and dust cleaning.

The raw materials are fed into the appropriate divisions of the storage system by autotransport from where they are charged into storage bunkers 2, 3, and 4 by an overhead crane or conveyors. Sulphur and the modifying additives from the storage bunkers enter reactor 5 for preparing the modified sulphur binder while sands and aggregates enter drying drum 6 for drying and heating and travel on heated elevator 7 at a temperature of 140–145°C to heat-insulated bunker 8, while the mineral flour goes to bunker 9 through feeding unit 10. With the help of appropriate dosing devices 13, the required amounts of mineral flour, sand, and aggregate are charged into heated concrete mixer 14, the dry mix mixed for 1–1.5 min after which the modified sulphur binder is added to the concrete mixer from reactor 5 through doser 13. The mix is again mixed for 2–3 min. The prepared mix is charged directly or with the help of concrete layer 15 into the heated mould in which the required reinforcement and inserts have already been arranged.

Vibration moulding is carried out on vibration platform 16 until the appearance of a liquid phase (sulphur paste) on the product surface but for not more than 2–3 min. The moulded product or structure together with the mould goes for striking. Striking and control are effected after the cooling of the article to a temperature of 35–40°C. After striking, the moulds are returned to the shops

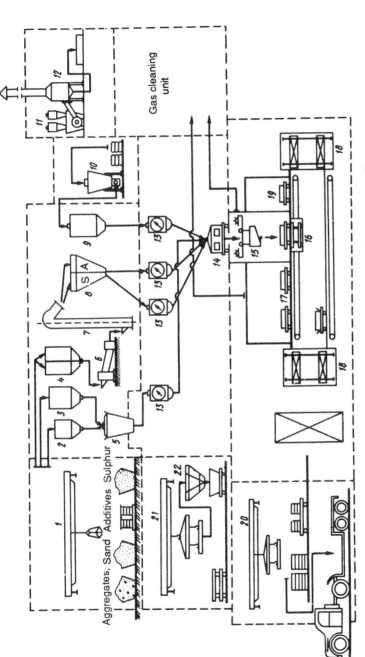

Fig. 10.12. Technological flow sheet for producing polymer sulphur concrete products:

1—storage of materials with an overhead crane; 2, 3, 4—storage bunkers for modifying additives, sulphur, sand, and aggregates; 5—reactor for sulphur modification; 6—drying drum; 7—elevator; 8—bunker for sand and aggregates; 9—bunker for mineral flour; 10—unit for feeding mineral flour; 11—cyclone; 12—dust collector; 13—dosers; 14—heated concrete mixer; 15—concrete layer; 16—vibration platform; 17—trolley with moulds; 18—trolley conveyors; 19—chamber for heating moulds; 20—storage of finished product; 21—section for processing rejects; 22—jaw crusher.

for cleaning and preparation for the next moulding cycle while the product goes to storage.

The section for processing the rejects is provided with a high-power press (not shown in the flow sheet), jaw crusher 22 and overhead crane.

The gas and dust cleaning section is provided with cyclone 11, dust collector 12, and devices for cleaning and neutralising the gases free of SO_2 and for catalytic burning of H_2S.

The production of sulphur concretes provides for the required mechanisation and automatic control of the entire technological process from a central control panel.

11

Rational Applications of Polymer Concretes

11.1. Use of Polymer Concretes in Residential and other Civil Constructions

When used in residential and other civil constructions, polymer concretes should primarily possess high decorative and aesthetic qualities, atmospheric stability and abrasion resistance, and to a lesser extent, resistance to acidic solutions and other aggressive products, which are more characteristic of industrial establishments.

On this basis, special requirements are prescribed for aggregates, monomers and oligomers. Thus, furan resins are black in colour and there is no possibility of producing polymer concretes in a bright range of colours using these resins. The use of basalt and other mineral rocks of black colouration as fillers and aggregates also renders difficult the production of high-quality decorative polymer concretes. In some cases, even an incorrectly selected hardening system may diminish the anticipated decorative effect.

When producing decorative polymer plaster or polymer concrete with special properties, particular attention should be paid to the selection of aggregates.

1. When producing polymer (artificial) marble products, not only the level of technical properties of the material, but (more importantly) also the attractiveness of external form is of decisive importance. The formula of wall thickness adopted in many cases is of no relevance in the case of coarse-grained fractions since it is absolutely impossible to use fractions with a granule size about 6 mm for producing polymer marble even if the average wall thickness is about 20 mm. This is due to the following reasons:

—When producing polymer marble, the plaster is pigmented with dyes; it is important to create a uniform marble texture by adding the dye; the texture should create a definite aesthetic effect and the individual fractions of aggregates should not be visible; the larger the size of individual aggregate granules, the greater the danger of their standing out on the material surface, especially when the main colour selected for the plaster differs from the natural colour of the aggregates; often, light shades are selected for artificial marble but aggregates of brown or grey shades are more readily available and hence problems of matching colours arise quite often;

—Colour formation imitating the marble shades is effected in the course of distribution of the plaster mass in the mould; it is therefore important to prepare the polymer plaster in such a consistency that it can be cast into the mould quite uniformly with the dye mixed well with the synthetic resin to ensure good masking of the filler;

—To produce plasters of polymer marble, aggregates with a maximum grain size of up to 1 mm can be used; the following indices serve as a rough guide: 50% of fine fractions (0.5–0.7 mm or 0.5–1 mm) and 50% of extremely fine fractions (less than 0.1 mm).

2. It is impossible to produce the so-called 'onyx effect', i.e., translucent or non-transparent polymer plaster, using sand fillers. The production of polymer onyx requires the use of fine-grained fillers possessing light refraction indices proximate to those of the resins used. In practice, aluminium hydroxide or frits are used for this purpose. Both these materials have a grain size of less than 0.1 mm and hence in such cases fillers with a grain size up to 0.1 mm are used.

3. Other types of polymer plasters (for example, imitation granite) are produced by using aggregates of different shades in combination with colourless resins. The aggregate should be selected on the basis of the required outer form of the product surface and not on the basis of the principles adopted when producing loose materials.

4. It is very important to remember that polymer concretes or polymer plasters resemble stone on increasing the content of the aggregates; contrarily, they more resemble plastics when significant amounts of binders are present in the composition. Properties, such as high compressive strength and deformation resistance call for the use of significant amounts of aggregates, while high bending, tensile and impact strengths necessitate the use of a high binder content.

When making boundary columns with a wall thickness of 5 cm (with no aesthetic requirements whatsoever), coarse-grained fillers can be used under the condition that the impact strength requirement is not particularly high. To ensure high impact strength, for example, of a pile to be driven into the ground, the resin content of the polymer concrete should be increased. The required high content of resins can be ensured only by increasing the total surface area of all the aggregates, i.e., very fine granules should be added to the composition.

Altogether new problems associated with the elimination of wastes by recirculating them are presently being faced. Thus, in the natural building stones industry, for example, in the field of mining of natural marble or granite processing, wastes constitute a significant quantity: these are in the form of large and fine stones, sands and finely ground fillers which, with no particular difficulty can be used for producing polymer marble. A new material called 'aglomramor' (marble agglomerate) has been produced from such waste material, primarily in countries in which the production of natural building stones is quite high. What is more important is not the search for the most optimum aggregates as the utilisation of the available aggregates for producing material of high quality so that, on the one hand, the waste materials are utilised and, on the other, become remunerative. It is known that 'aglomramor' in spite of being very low in quality compared to polymer marble, and with a less aesthetic external form offers high economic advantages and hence its use is extremely attractive.

Like any other building materials, polymer concrete should be used at places where its excellent properties can be put to maximum use (high mechanical strength, chemical resistance, water impermeability, diversity of forms, etc.) and, at the same time, its comparatively high cost compared to other materials based on cement binders justified.

This possibility has been demonstrated successfully for several years by plants producing concrete products and structural members, which have taken to producing polymer concrete partly or fully or set up technological units for producing polymer concrete along with cement concrete products. These companies use polymer concrete for producing technical components and products taking advantage exclusively of the high quality and excellent properties of this material. The external appearance of the product is determined by the fillers used and their size composition since the binders, with the exception of some (furan resin, for example), are colourless and transparent.

The cost of producing polymer concrete products is somewhat higher than cement concrete products designed for comparable application or load. This is not surprising as polymer concrete products are of superior quality compared to the cement concrete products. In spite of the fact that the cost of the raw materials used for producing polymer concrete products is eight times more than those used for producing cement concrete products, this difference decreases subsequently in the course of production with the result that, ultimately, the cost of polymer concrete products is only 10–25% more than that of comparable cement concrete products.

This economy has become possible because about 80% of material economy is achieved by reducing the overall sizes of the products thanks to the high strength of the polymer concrete; the transport of materials and products is greatly reduced because of their smaller bulk; capital expenses are considerably reduced on the hoisting mechanisms and other ancillary operations; the process of producing the finished products is accelerated as a result of rapid striking; and the area required for producing and stocking the products is less because of the rapid hardening of the material.

The above advantages of the polymer concrete industry have been particularly effectively utilised by firms and plants which have already specialised in producing synthetic resins and their products, i.e., those which possess some experience in this field and thus can readily find new customers for their products.

Nevertheless, many companies and plants having neither production experience nor demand have embarked on the polymer concrete industry. Such firms have begun the search for new possibilities, quite remote from their earlier activity, which ceased bringing in adequate returns as a result of stiff competition or for other technical reasons.

The diverse possibilities which polymer concretes and their products possess offer good and highly promising prospects to all those who are interested in producing them.

The client readily accepts a certain increase in product cost since it is compensated by a much higher product quality and the possibility (to the buyer) of economy by reduced expense on the transport of the products to the construction site, quick assembly, high accuracy of sizes and dimensions of products, and their easy manoeuvrability.

Thus, works traditionally producing cement concrete products and precast structural members use polymer concrete to meet the requirements of their regular customers for articles of very high quality and also for introducing greater diversity in their production programme.

Polymer concrete is an excellent alternative to natural stone in the form of slabs or panels (i.e., for making wall panels, window sills, staircase steps, floor tiles, table tops, etc.) and also for ceramics in the field of producing sanitary engineering goods.

11.2. Polymer Concrete in Construction of Underground and Engineering Structures

The most popular use of polymer concrete is in various drainage systems and canal networks: 'AKO-drain', 'Polydrain', 'Polycast', etc. These systems meet the current engineering requirements and are produced and used in various countries. The system consists mainly of channels and shafts joined by extremely diverse methods. Fig. 11.1 depicts the various members and the possibilities of joining them for making composite drainage and canal systems.

Polymer concrete members are made to a wall thickness of 15 mm enabling a weight reduction to one-third of similar products made of ordinary concretes using cement binder. The slope required for water flow is incorporated in the design of the channels and hence they can be laid on the surface with no additional slope. The in-built slope in the channel design, the semi-circular form of the bottom and the very smooth surface of the channels ensure a 10-fold greater rate of water flow compared to the ordinary drainage systems.

As a result of low water absorption, polymer concrete possesses exceptionally high resistance to cold and thaw. Field tests on freezing and thawing showed that cement concrete lost about 25% of its initial weight after 750 cycles while polymer concrete maintained its initial weight even after 1600 cycles.

The high resistance of polymer concrete to the action of chemical agents

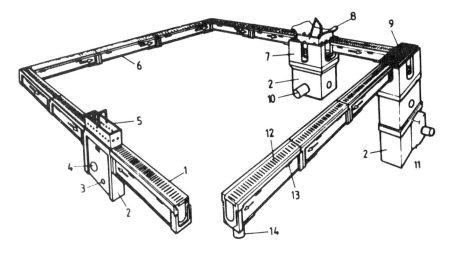

Fig. 11.1. Drainage system:

1—steel grating; 2—collector; 3—small diameter opening for joining the overflow pipe (hose); 4—large diameter opening for joining the overflow pipe (hose); 5—slimes container (settler); 6—channel with a dip of 0.6%; 7—inlet of channels; 8—slimes separator; 9—frame and grating of cast iron; 10—overflow; 11—siphon; 12—cast iron grating; 13—device for fixing the grating; 14—overflow.

makes it suitable not only for the transport of water contaminated with chemicals, but also for laying drainage systems in chemically aggressive soils, as for example, in soils containing humic acid, with no possibility of damage in the course of prolonged use. Thus, the service life of drainage systems made of polymer concrete can be regarded as unlimited.

Depending on the type of soil and other requirements, the channels are laid straightaway on the ground or are protected from lateral sliding by a bed of solid concrete (Fig. 11.2).

When selecting covers, it is important to ensure the required strength conforming to the anticipated load and the dimensions of the inlet section matching with the water flow required to be transported and the site of laying of the system (roads, aerodromes, automobile parking lots, garages, footpath zones, sports fields, basins, etc.). Such covers in most cases are made of stamped metal or cast iron (Fig. 11.3).

Under extreme loads, for heavy duty channels, covers of cast iron firmly fixed to the channel are used. The appropriate fixing components are cast directly in the course of producing the channels.

A Dutch concrete plant produces 200,000 rain water drainage box units of polymer concrete for use in the streets. This plant used to produce such box

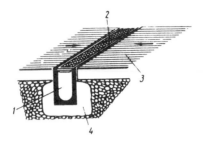

Fig. 11.2. Sketch showing laying of polymer concrete channels:

1—polymer concrete channel; 2—channel cover of expanded metal; 3—flooring; 4—monolithic concrete about 8 cm thick.

Fig. 11.3. Types of covers for channels:

1—grating of expanded metal or loop grating with cell or loop dimensions of 10/30, maximum wheel load 6 tonnes, total load up to 25 tonnes, inlet section about 700 cm^2/m*; 2—heavy-cast grating, maximum wheel load 10 tonned, total load up to 60 tonnes, inlet section about 215 cm^2/sec*; 3—grating of polyethylene set flush, maximum wheel load 1 tonne, total load up to 3 tonnes, inlet section about 190 cm^2/m*; 4—grating of polyethylene in the form of a seat, maximum wheel load 10 tonnes, total load up to 60 tonnes, inlet section about 152 cm^2/m^2*; 5—lids for sports fields in the form of a saddle especially designed for sports fields.

units out of concrete using a cement binder to a wall thickness of about 5 cm. The discharge pipe and grating were formerly made of cast iron in a separate technological process. At present, these metal components are laid in a mould and polymer concrete cast in it. The residence (stay-in) time in the mould (i.e., the time required for filling the mould, hardening of the polymer concrete, withdrawing the core, and striking) is 6 min. At the end of this period, the finished product can be transferred to the rack and after 2 h is ready for despatch. In the Benelux countries and in Japan, post offices use distribution boxes made of polymer concrete for underground use.

In the delivery and receiving reservoirs of cleaning systems, diversion channels are provided for water transport. In most cases, these are made of anodised

---

* Units so given in Russian text—General Editor.

steel, aluminium or stainless steel.

Compared to the concrete channels (made of reinforced, light-weight concrete or polymer concrete), metal channels cost twice (aluminium grade 57S) or thrice more (steel with a protective coating or stainless steel). It is impossible to produce these members using asbestos cement.

The following are the advantages and disadvantages of concrete structures.

Ordinary concrete with gravel aggregate. Structures made of concrete are dependable but they are quite heavy. As a result, very heavy reinforcement has to be used for making cantilevers and walls (compared to metal channels). Further, problems can arise when drawing tension cables.

Light-weight concrete. When using light-weight aggregates, the weight of channels decreases to 70% of the weight of ordinary concrete. There is then no need for additional reinforcement of cantilevers and walls. As a result of weight reduction, assembly becomes easy. A drawback of light concrete is the more intense, compared to gravel-based concrete, crack formation, which is visible in facades of thickness 10 cm.

Polymer concrete. As a result of the high strength of the material, the weight of a channel made of polymer concrete is no more than one-half that of a corresponding member made of heavy concrete or about two-thirds of members made of light-weight concrete. Polymer concrete is extremely stable in chemical media and is cold resistant because of its low water absorption. Its surface is smoother than that of cement concrete and hence reduces the danger of atmospheric precipitations settling on it. Channel elements made of polymer concrete can be produced to high dimensional accuracy and are easily replaced later.

Polymer concrete is highly recommended in many fields of construction. It has been used for building chemical plants, for example, in fabricating electrolytic equipment. Reservoirs, shafts, support structures, etc. are now being produced from polymer concrete.

The reason for this shift to polymer concrete is that the materials used before, such as cement concrete with artificial reinforcement or wood with special impregnation no longer meet the growing requirements for ensuring chemical stability. Moreover, the high mechanical strength of polymer concrete offers advantages with respect to design and costs.

Large-sized polymer concrete members can be made as individual components to be assembled later into the required structure or as a monolith. The type of production depends on technical as also transport and assembling conditions. When a member is made as a monolith (for example, oil separators), it should be made as far as possible without need for any additional processing. When producing as individual members for assembly into large units or blocks, for example, foundations of machinery or transformers, it would be advantageous to so design the mould as to economise on the subsequent finishing operations. Thus, it would be preferable to introduce threaded inserts in empty moulds and later fill them with polymer concrete rather than to bond these inserts in holes drilled later after producing the member: firstly, the labour in the first method is less and secondly the inserts hold better in the material.

A good example demonstrating that polymer concrete and cement concrete need not necessarily be viewed as irreconcilable competitors is provided by their use in Switzerland for making drainage channels in railroad tunnels. The aggressiveness of the water and the difficulty and expense of cleaning the channels made imperative selection of the most appropriate material before undertaking any drainage system. Channels made of traditional materials were unsuitable because of their porosity and low surface hardness. A combination of them with polyvinyl chloride shells was also tested. However, the quality of the surfaces did not meet the stipulated specifications. The possibility of using finished concrete channels with an improved surface quality was examined. Engineers of the Federal Railroad of Switzerland designed in collaboration with specialist-chemists composite members of reinforced cement concrete which were used together with polymer concrete shells. These conformed to the special requirements of chemical stability, thermal stability and mechanical strength.

The following are the additional advantages of such shells: long length of 4 m each without danger of distortion or deformation of members, high dimensional accuracy of ± 1 mm and pigmentation of polymer concrete. A black pigment was added to the polymer concrete for visual identification of lime sediments.

The ready-made polymer concrete shells were placed on the bed of the channels and filled with cement concrete. Adhesion to the reinforced cement concrete was ensured by the roughness and form of the rear side of the shell (in the form of a dovetail). The six-tonne channels were assembled using specially made packing devices. The joints between the individual segments were joined using grooves and pins and a special hermetic sealing device. During this work, continuous rail traffic, partly on one track, plied on the international railroad section between Italy and the FRG.

No perceptible lime depositions were seen on the smooth and nearly nonporous polymer concrete shells during the four years of observations in this experimental section.

11.3. Use of Polymer Concrete for Covering Facades

Facade designs used to be largely determined by the specific features of the building materials used in a given locality. Thus, framework facades with overlaps sealed by clinker aggregates and facing with natural stone or plastering were mainly practised. Further, plaster work was provided with additional protection from atmospheric effects by wooden or slate shingles. The proportion of the facing with an air layer in-between the interior and exterior wall layers was relatively small.

In recent decades the building industry has been enriched with a wide range of facing materials which have influenced the design of façades. Now attention has focussed on the structural and physical properties of materials when designing buildings and installations.

An architect is often faced with the problem of intelligently combining external appearance with optimum structural design without overlooking the cost to the builder, while selecting materials for facades from the wide range of such available in the market. In the immediate future, while making such decisions, he will also have to satisfy new regulations on power economy. Polymer concrete has become an alternative high-strength building material. This is of particular relevance when fire safety has to be ensured. At present, the materials used for facades have to meet the requirements of 'not readily inflammable' or 'self-extinguishing' classes. These requirements are fulfilled by using small amounts (less than 10%) of binder to the polymer concrete with no other additive. But when the binder content is very high (for use as decorative facade material, this is an essential condition), fire safety requirements are fulfilled by adding antipyrenes, for example, antimony trioxide, which permits using polymer concrete as a material for facades of tall domes.

The dimensions or the external appearance of polymer concrete panels (tiles) used for facing purposes are practically unlimited. Panels imitating natural materials (slate, granite, basalt, etc.) and large-sized products, for example, tiles in the shape of shingles, can be produced in polymer concrete. The thickness of the tiles and the method of their affixture on the facade depend on their weight.

Facing tiles are laid predominantly in the form of suspended members with an air (ventilation) gap and an additional insulation layer. Large-sized sandwich-type members are also produced; in these, the middle insulation layer (polyurethane foam) is cast in the course of production itself.

Let us give some examples of facade components made of polyester concrete produced at one of the factories in the Netherlands with a wall thickness of 15 mm, panel weight of about 25 kg/m^2 and size of panel up to 3 m^2; panels of area up to 6 m^2 are 20 mm thick and weight 45 kg/m^2. These panels can be used with good economy in repair and renovation work.

Most types of suspended facades made with an air ventilation layer are affixed to a wooden or metallic base. It is possible to do away with this base, however (Fig. 11.4).

The facade element is affixed directly to the building wall (using dowels) with anchors of alloy steel prefixed to the panel at the time of its casting. When using additional insulation material, appropriate holding devices are provided in the panel. The anchors are flexible and are bent at the time of assembly to match with the required thickness of the ventilation layer. The passage around

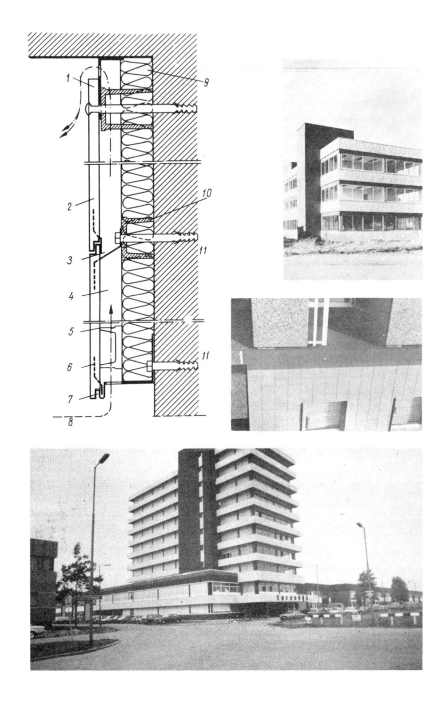

the element is in the form of a joint which offers protection from rain and ensures compensation of atmospheric temperature variations. The assembly is simple and fast.

When the architect is desirous of producing distinctive individual components, polymer concrete members can be combined with other materials and products available in the building materials market.

If the heat insulation indices of the outer walls of the building or installation do not match with the required specifications, an additional insulation layer (made of ordinary thermal insulation material) can be affixed between the covering and the outer wall through an insert made of material resistant to microorganisms. This enables the use of polyester covering elements for all types of work on buildings. The optimal field of application of these products should be arrived at while keeping in mind the requirements of power economy for repair and renovation operations or modernisation (reconstruction) of old and damaged building facades.

For affixing the elements to the outer walls, the dowels available in the market and endorsed by the building inspectorate may be used.

In addition to facades of different types, attics, balconies and devices for landscaping can also be made of polymer concrete. Thus the finished building becomes an 'architectural landscape.'

Decorative facade elements of polymer concrete are generally made with a decorative top layer consisting exclusively of the binder with the addition of pigments, i.e., without aggregates. This outer decorative layer ensures a close filling of surface voids in the polymer concrete. This prevents the formation of hollows and fading of the polymer plaster. Moreover, the presence of this layer intensifies the surface shine on the decorative elements. When coloured elements are required, the pigment can be added either to the plaster or the decorative layer. To produce single-coloured members, pigment is added to the decorative layer; contrarily, to produce, for example artificial marble, the decorative layer is left transparent to emphasise the presence of diverse shades in the polymer mass. The transparent decorative layer is laid on the surface of polymer concrete elements and, when it is required to create special external effects, multicoloured fillers are added as, for example, when producing imitation granite.

Fig. 11.4. Facade cover and various types of facades:

1—ventilation profile; 2—imitation slate made of polyester; 3—upper brace for mounting; 4—heat insulation layer; 5—insert for ensuring ventilation between the covering and the wall; 6—lower brace for mounting; 7—combination profile (support); 8—direction of ventilation space between the covering and the wall; 9—glass wool insulation; 10—insert; 11—dowel for screw with a hexagonal head.

11.4. Decorative Polymer Plasters for Interior Finishing

Polymer plaster can be successfully used for finishing the interiors of buildings and is encountered in extremely diverse forms. Apart from the excellent technical properties of the material, the main reason for its extensive application for interior finishing is the diversity of decorative properties of polymer concrete. Often, polymer plaster is used for producing imitation marble when it is compared (in cost and quality) with the natural marble. It is known that suppliers are compelled to stock large quantities of natural marble in warehouses to be able to meet all categories of requirements. Natural marble is highly brittle which adversely affects the material transport and processing operations and ultimately its cost. Moreover, marble cut from blocks is available in the market only as tiles of limited sizes as the material does not possess the property of transverse tensile strength. Panels beyond 2 m in length and 0.5 m in width are usually not available. Bent or curved forms are not available at all (for example, for use in producing sanitary goods).

When used in the form of window sills, table-tops, staircases, or tiles for covering floors, the material is exposed to the action of water flowing from flower pots, spilled drinks, dirty water, floor cleaners, abrasive loads, atmospheric effects, etc. These usually leave traces on the material surface. For these reasons, the marble surface is protected with a transparent layer of unsaturated polyester.

However, in spite of all these adverse factors, marble continues to be used and, in fact, is known from ancient times as a valuable material for providing good interior decorations. Thus, it is not surprising that attempts are made to take advantage of all the possibilities offered by its artificial counterpart, i.e., polymer marble. The beauty of colour shades of natural marble is fully captured while at the same time the technical properties of the material are greatly improved.

Polymer marble represents a special type of polymer concrete. Practically, the very same binders and fillers are used except for much finer sizes of granules. The effect of the outer form of natural marble is created by adding pigments of the required type. The fine sizes of filler fractions and the resultant higher content of the binder impart to polymer marble a much higher transverse tensile strength, impact strength and tensile strength compared to standard polymer concrete.

The initial response to artificial (polymer) marble was adverse, as in the case of many other new materials. Adverse phenomena associated with the new product were highlighted. Those selling the licenses attempted to sell some formula for a certain fee and charged fancy prices for additional supplies of materials. Many who bought the licenses soon realised they had been duped. Subsequently, however, as a result of work in the neighbouring disciplines (established chemical concerns supplied binders while engineering workshops supplied equipment for producing polymer marble), it became possible to establish the production of high-quality polymer marble. It should be recalled in this context that many industrial and experimental flow sheets are available to ensure stable and high-quality material. Special mention should be made of the contribution of the Cultured Marble Institute (USA) which has formulated comprehensive instructions and recommendations for the USA, which produces the world's largest volume of polymer marble.

By selecting the most suitable raw material (special attention should be paid to the light stability of the pigments used; in the FRG, dyes commencing class 8 on the local Woll's scale are considered suitable) and utilising effective mixers and dosers, non-porous plaster surfaces, uniform pigmentation and internal structure have become possible and fissure formation, warping and lamination eliminated; in other words, a high-quality product is ensured.

At present, polymer marble panels are produced to standard dimensions: length up to 3500 mm, width up to 600 mm and thickness up to 20 mm. Such panels can be cut to the dimensions required which helps the suppliers to fill orders quickly, say for window sills. There is no danger at all of crumbling or chipping of the material while cutting. The surface of the cut edges is even and smooth and the technique adopted enables cutting the material such that the cut edge completely matches with the edge of the panel. Buyers not desiring to stock a large amount of the material can obtain finished goods cast to the given overall dimensions. In such cases, the buyer has no need to cut the material at all but the time taken for supply correspondingly increases.

Transverse tensile strength (25 N/mm^2) permits transporting panels 3500 mm × 600 mm (thickness up to 20 mm) by hoisting on either side without fear of breakage. Polymer concrete panels of length 3500 mm imitating marble can be carried on staircases to the point of placement without damage. The decorator should master the finishing of polymer by cutting and drilling or assembling them by using plasters or adhesives.

Housewives have a particular preference for material whose surface can be made lustrous or mat and whose maintenance is easy and even pleasant. Dirt settling on polymer concrete surface can be simply washed with water with no additional treatment (for example, toxic fluosilicates).

Polymer concrete panels are also used for making staircase steps and for this purpose their surface is made slightly lustrous-mat and shaped. Wear tests using the Beme disc conforming to the FRG standard DIN 52108 gave an abradability value of about 5 cm^3 per 50 cm^2 test surface after 440 revolutions of the disc; in other words, these parameters are much superior to the corresponding mechanical properties of natural marble.

A new type of polymer marble, i.e. polymer onyx, possesses a certain amount of transparency and surface excellence like that of marble but has a much greater depth effect. To ensure the effect of translucence of the material, the refractory index of the filler should be similar to that of the binder. At present, the market offers a wide range of extremely diverse fillers but aluminium hydroxide is most often used. This powder is produced industrially in various types and qualities. The cleaner (whiter) the material produced, the higher the expense involved and hence the greater its cost. However, it is not often that a high grade of the material is required. Beige or brown shades are quite acceptable. The buyer accepts this grade of powder as wholly suitable since the use of materials with a brown shade in residential buildings makes for a feeling of warmth and the architects take advantage of this phenomenon.

When comparing polymer marble with polymer onyx, it should be remembered that the material imitating onyx is produced in the same manner but is somewhat more expensive. This is because, on the one hand, the filler itself is expensive and, on the other, the content of the expensive binder in the plaster is more.

Extensive use of polymer marble and onyx began with the availability in the market of moulds for producing tubs, sinks, toilet bowls, bathroom floor fixtures, cans, etc., which made possible the commercial production of sanitary goods.

In the production of sanitary goods, one of the most important and decisive properties of polymer marble is its water impermeability, especially on frequent changes of cold and hot water. The Cultured Marble Institute (USA), designed the following method for testing this property: a bowl is filled with hot water at about $66 \pm 5°C$ and held for 1.5 min. This water is drained in 30 sec and replaced with cold water at $10 \pm 1.5°C$. This cycle is continuously repeated 500 times and then the surface of the material examined. The formation of cracks or chipping of the product is not permitted. Among the other American methods for testing polymer materials is the test for strength and colour fastness conforming to ASTM standards and also the special methods developed by the CMI for testing the properties of artificial marble: resistance to oil spots, chemical resistance, abrasion resistance, possibility of cleaning and washing the products, and the so-called 'burning cigarette test' in which a burning cigarette is allowed to rest on the surface of the polymer marble or onyx for 2 min. This should cause no inflammation of the material and the brown spot formed should be easily wiped off or removed by rubbing with fine emery paper.

The American methods of testing the materials were developed primarily keeping in view the extensive use of polymer marble sanitary goods in hotels for which the quality requirements are particularly rigid, far more rigid than for use in residential buildings in which the sanitary articles are regularly used by the same persons and their adequate care is assured.

Apart from using it for making sanitary goods, polymer marble or onyx is now used far more extensively in kitchens for facing the walls and making table tops, often using material of one single colour. The main reasons for such extensive use is the possibility of making diverse shapes and tiles of required dimensions, resistance to the action of aggressive acids used in the households (citric and lactic etc.) and also the desire to avoid plastics in households and use beautiful 'stone' instead.

The diversity of possible shapes and atmospheric stability constitute the basis for the extensive use of the material for producing statues, decorative vases, and other small architectural forms as well as tombstones (Fig. 11.5). Numerous articles can be produced using a single mould well imitating the natural stone material with no chiselling whatsoever. Thus, labour and monetary economy is ensured and the growing environmental conservation requirements are met as the use of artificial material makes for less intense utilisation of natural stone.

Fig. 11.5. Statues and small architectural forms made of polymer concrete.

11.5. Use of Polymer Concrete for Covering Floors and Roads

Polymer concrete and polymer plasters have long been used for covering standard cement floors. Rapid setting facilitates the use of polymer plaster for repairing road and other concrete surfaces.

While carrying out repairs, it is important not to close the road to traffic any more than necessary. For this purpose, mobile make-shift construction groups are organised and the road closed to traffic in parts. Material, machinery and workers are transported on trucks from one site of repair to another.

After carrying out repairs in a section in 1963, the covered surface underwent a 100-fold treatment with salt solution or was scattered with sand and fine rubble for over a year. In 1964, it was established that only 3% of the surface was damaged in patches. Often, the plaster for repair was laid together with concrete which the plaster held firmly and the concrete was no longer damaged during further repairs. On 1 m^2 of patch, about 10 kg of polymer material were used, including levelling of depressions (up to 15 mm depth).

From patchwork, covering of large surfaces of roads and sometimes even whole streets or roads was taken up. Investigations gave divergent results. For example, there were repaired sections whose surface quality remained high even after several years of use and the repair could be regarded as wholly satisfactory. But there were others whose surface condition was not of sufficiently good quality even after a relatively short period. Such an uneven quality of repair was not so much due to the properties of the materials used or their formulae as to the working methods, a judgment derived from the significant differences in the behaviour of absolutely identical materials and formulae.

Special attention should be paid to the good quality and soundness of preparation of the surface on which patchwork is required. If, for example, the surface has not been fully cleansed from contamination and defective concrete lumps have not been removed, the coating will not adhere well to the concrete base and the polymer concrete will peel from the road surface after a brief spell. To ensure good adhesion, it is also important to do good priming of the surface; appropriate equipment is essential to ensure uniform and quality mixing of the plaster; weather, too, plays a significant role when carrying out repairs.

Inadequate observance of directives and norms of production technology (for reasons of urgency or due to lack of practical experience) will lead to innumerable defects and to the conclusion that large-scale use of this technology is premature and that further improvements are necessary.

Results in covering concrete surfaces in closed buildings have been quite different. There were no problems associated with atmospheric action nor with work quality as the job was usually carried out leisurely by experienced workers. Hence positive results attributable to the excellent physical and chemical properties of the material are not surprising.

Floors in the production areas of industrial establishments are constantly exposed to intense loads. For example, transport media as also vibrations, impacts, and abrasion exert constant mechanical action on the floor surface. Action of chemicals (acids, alkalis, fats, oils and salts) also affects floor quality.

Under constant loads and other effects, unprotected floor surfaces are soon damaged and pose danger to the health of workers and adversely influence the state of the machinery. Continuous process operation becomes impossible leading to losses; the cost of repairs is then extremely high.

To prevent serious damage to floors exposed to constant mechanical and chemical actions, it is not sufficient to strengthen them here and there. A full application of coverings prepared from reactive resins (without solvents) containing a filler and pigment offers protection against damage.

As in all other areas of the application of polymer concrete, here, too, the correct selection of binder is important. It is hardly possible, however, to regard any one particular raw material as optimal and the others as unsuitable. As already pointed out, all materials have specific advantages and disadvantages. In western Europe, methacrylate, epoxide, and polyurethane resins (in some rare cases, polyester resins) are mainly used in making floor coverings.

The successes achieved in laying protective coatings in closed buildings, further developments in binders, and also a new generation of engineers have renewed the work on developing protective coatings even for objects exposed to the open sky. The forerunners in this field are specialists of the USA, a country in which the extent of automobile roadway was for long more important than its quality.

One of the most serious problems facing the US highway engineers, in fact engineers everywhere, is the rapid damage of concrete coverings based on portland cement as binder. These were used for constructing bridges, flyovers, and highways. The porous concrete is open to the penetration of water and other liquids which react with the concrete and, as a result, chlorides from salt solutions used for thawing the ice enter the concrete. This accelerates the corrosion of the reinforcement. Corrosion enlarges the volume of the steel bars and generates internal stresses in concrete which ultimately cause the formation of cracks and breakdown.

In regions with sharp temperature variations, water penetrating into concrete freezes and expands and the resultant stresses break up the concrete. Intense movement (primarily of heavy trucks) greatly reduces the chances of timely braking of the vehicles and thus causes intense wear of the concrete surfaces, leading to increased danger of road transport accidents. An improved quality of concrete should be considered in this context just as necessary as its strength and durability.

The need for applying coverings ensuring protection from slippage, penetration of water and chlorides into the concrete, and also the restoration of stability against slippage in the case of worn-out concrete surfaces has long been recognized. Most promising is the method of laying thin polymer concrete coverings which form a protective barrier against the penetration of water, salts and other chemicals, and ensure effective durability against slippage for a long period.

The results of investigations carried out for ten years in the laboratory and at test sites indicate that the efficiency of thin polymer concrete coatings depends on the properties of the resins. An appropriate analysis helped in determining the thermal tangential (shear) stresses arising in the zone of contact of the polymer concrete covering with the surface of portland cement due to differences in the physical properties of these materials. Thermal tangential (shear) stresses were calculated using van Vlak's equation:

$$S = (\alpha_p - \alpha_c)E_p E_c dT/(E_p + E_c), \qquad \ldots (11.1)$$

where S is tangential stress; α_p and α_c coefficients of thermal expansion respectively of polymer concrete layer and concrete based on portland cement binder; E_p and E_c the modulus of elasticity of the polymer concrete covering and of the portland cement concrete; and dT the temperature variations.

It is important to remember that thermal tangential stresses can be quite high if the coefficient of thermal expansion of polymer concrete is compared with the corresponding index for portland cement concrete. Experience has shown that the tangential stress should not exceed 0.03 $N/mm^2 \times °C$ to ensure a life of not less than ten years for the covering.

In view of the fact that the coefficient of temperature expansion of polymer concrete of all grades in general exceeds the corresponding index for portland cement by 2–5 times, the modulus of elasticity of the polymer concrete should be held at as low a level as possible. Moreover, the resin should have an expansion exceeding 30% in order to compensate for the internal stresses arising out of temperature changes. It is recommended that the tensile strength of the resin shall be more than 200 psi [sic] while the modulus of elasticity may not be less than 200,000 psi.

Further, the resin should not come into contact with the aggressive alkaline bases and, as far as possible, polymer concrete should not be laid on a wet concrete surface. This is permissible only when the resin and the hardener are not sensitive to the action of moisture.

Resins with the following characteristics are recommended for use in polymer concrete coverings: elongation 30 to 50%, tensile strength not less than 14 N/mm^2 and modulus of elasticity not more than 1400 N/mm^2.

The well-known example of the successful use of polymer concrete for protecting road surfaces is the repair of the Brooklyn bridge in New York designed so wide a century ago that it now handles six-lane automobile traffic.

Innumerable investigations and tests of different materials for laying road coverings of various thicknesses led to the selection of epoxide resin with besalt as filler for this purpose. The thickness of the layer applied is 12 mm. In the first section of the bridge 110 tonnes of polymer plaster were laid in 42 h. Four hours after laying, the covering could take the full traffic load. Work was carried out lane by lane so that the traffic on the bridge was never completely shut down.

New areas for the application of polymer concrete coverings have been identified. Thus, polymer concrete coverings are being laid on large surfaces of water transport channels to prevent cavitation at high rates of water flow and thus protect the concrete surfaces from damage.

Polymer concrete with good light filler-insulator has been used in covering the surface of liquefied gas storage collectors. With this covering, the thermal conductivity of concrete collector tanks decreased to one-fifth its original level. This has greatly prolonged the duration of evaporation of liquid gas in the event of a possible blow-out.

11.6. Polymer Concrete in Agriculture and Horticulture

In many countries, polymer concrete is being used for making feed troughs for use in piggeries and cattle farms. The main reasons for using polymer concrete in this field are the strength and low weight of polymer concrete members compared to the earlier ones based on cement binder, its superior chemical resistance to the action of ammonia vapours common to all animal farms and action of synthetic detergents used in removing pathogenic bacteria to prevent contamination of feed. Compared to the troughs made of ceramics, polymer concrete ones are much cheaper and less prone to mechanical damage.

Polymer concrete can also be used for making floors with clearances for dung removal. Such floors possess high thermal capacity due to the low thermal conductivity of polymer concrete. This index can be increased by adding lightweight effective aggregates to the polymer concrete composition.

In horticulture, polymer concrete is used for producing troughs, tiles, small architectural forms and containers for growing seedlings. Such containers often have to be moved from place to place (depending on climatic conditions) and herein lies the clear advantage of polymer concrete—its low weight compared to other materials. Standard units 200 cm × 60 cm can be shifted without difficulty by two persons. Users gain a 50% economy in wages.

The other known fields of application of polymer concrete in agriculture are in making boundary and survey pillars in land management practice, as posts for fences, supports for vine creepers, etc. In well-laid parks, polymer concrete is being increasingly used for making benches and tables taking advantage of its durability. It is also used in laying playgrounds and sports fields, for making the tops of pingpong tables, tennis practising walls, etc.

11.7. Polymer Concrete as Sound Insulating Material

Polymer concrete is known as a good sound insulating material. This property is used for example, for setting up soundproof barriers on highways (especially on bridges and flyovers).

For producing sound insulating systems, the following are the specifications prescribed for the material: adequate strength, especially under intense wind load; resistance to the action of moisture, exhaust gases, detergents, engine oils, antifreezes, and also light; and weight restricted to 40 kg/m^2. Polymer concrete satisfies all these requirements.

The requirements are altogether different for dynamic loudspeakers. When using ordinary columns, made of wood until now, undesirable vibrations arise in the wooden structures. These vibrations are more intense when the panels and columns are fixed very rigidly.

Loudspeakers on polymer concrete columns minimise body vibrations. The body can be cast as a single piece and not assembled. With a high damping capacity, such a column even on intense loads at low (bass) frequencies does not experience vibrations and bass tones are heard distinctly and clearly.

One more advantage is the interference of sound waves at the edges of wooden columns. This phenomenon is unavoidable in the case of wooden articles since they are assembled. Polymer concrete helps produce columns of diverse shapes (spherical, tubular, or oval), thus preventing interferences.

It should be pointed out that, when making polymer concrete, apart from common (mineral) fillers, other material can also be used. There is at present a tendency to utilise material regarded as waste. For example, for laying the floors of industrial establishments and making the wheels of garbage cans and tanks, spent rubber (from old automobile tyres) can be used as filler in admixture with polyurethane resin. Such a product, strictly speaking, cannot be called polymer concrete but nevertheless is similar.

Glass represents another filler. By crushing waste glass, fine-grained inert powder can be produced which together with synthetic resin yields a combination glass-polymer suitable for making extremely diverse products. The utilisation of waste glass can, on the one hand, clear a large number of dumps and on the other, economise expensive natural material. Investigations carried out gave surprising results: strength of about 100 N/mm^2 (compressive) and strength on extreme compression up to 300 N/mm-m. These indices are wholly suitable at a wall thickness up to 25 mm and are thrice the values for cement concrete pipes. There were no adverse reports on chemical stability and impenetrability.

11.8. Use of Polymer Concrete in Industry

Load-bearing supports, i.e., props and tie-plates made of polymer concrete, for use in mine workings were made for the first time (in 1959) in the Institute of Mine Construction, USSR. The props were in the form of hollow posts of section 155 mm × 160 mm and length 2–3 m and reinforced with four longitudinal steel bars of diameter 12 mm with a crimped profile and cross-hoops (ties) in the form of a continuous spiral of cold-drawn wire of diameter 4 mm spaced 50 mm apart at the base and 100 mm along the rest of the length. To reduce weight, a right-through opening of diameter 110 mm was made in the prop (Fig. 11.6). When testing for central compression, breakage occurred under a load of 400 kN as a result of stress concentration in the end portions, i.e., props of this design possessed nearly the same load-bearing capacity as centrifugally-produced reinforced concrete members used for supporting rocks of moderate hardness while the weight of the new props was only half that of the earlier ones.

The upper tie-plates were designed with a T-section in lengths of 2300 to 2500 mm. They were provided with flat reinforced sections at the ends for resting on props. The longitudinal reinforcement was made of two rods of diameter

Fig. 11.6. Props for supporting mine workings.

16 mm with crimped profile. The use of much lighter props and tie-plates greatly promoted the working conditions and improved the labour productivity when driving galleries.

The experience gained at the Skuratov Works in producing steel reinforced polymer concrete structures helped in organising the commercial production of props and tie-plates. In a comparatively short period, over 21,000 units of such supports were produced. Over 10 km of mine workings are held with these supports in the Donetsk Basin mines.

The Dnepropetrovsk Engineering Design Institute designed vortical clarifiers using polymer concrete instead of conventional wear-resistant alloys for a cellulose-paper combine.

Feed bins made of reinforced polymer concrete have been successfully used at the Soligorsk Potassium Combine for several years. These bins, 3 m × 1 m and 2 m × 1.5 m have a wall thickness of 80 mm. The metal bins with a wall thickness 8 to 10 mm used earlier corroded in 6 to 10 months and became unfit.

Floor gratings for central sewages of cattle farms and industrial aggressive liquids (Fig. 11.7) are successfully being used in several agricultural and industrial establishments. In the state farms Ladozhsk and Dzerzhinsk in the Krasnodar region, two shops were built for producing 'warm' and chemical-resistant polymer concrete tiles for floors of cattle sheds. About 200,000 m^2 of such tiles were produced in these shops.

Good results were obtained in producing for a printing plant tiles using polymer concretes based on acetone-formaldehyde resins hardened in an alkali medium under the action of amine hardeners.

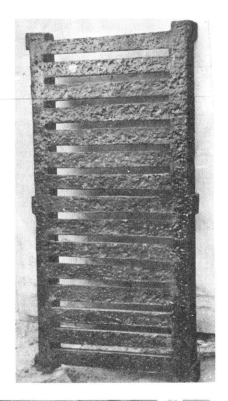

Fig. 11.7. Floor gratings for sewage channels and floors of cattle sheds.

The Reinforced Concrete Research Institute together with the State Institute for Designing Non-ferrous Metallurgical Plants and Voronezh Engineering and Design Institute designed, produced and tested an experimental set of load-bearing chemical-resistant structures made of steel-reinforced polymer concretes based on FAM resin for industrial buildings exposed to the intense action of liquid aggressive media (1968). Columns of platforms supporting baths (Fig. 11.8) of section 300 mm × 300 mm and length 35,000 mm had longitudinal reinforcements in the form of four rods of diameter 16 mm with a crimped profile and cross-hoops of cold-drawn wire of diameter 8 mm. Central compression tests carried out using a testing machine (capacity 12,000 kN) gave a breaking load of 4400 kN. On eccentric compression with an eccentricity of 75 mm, the breaking load was 2000 kN. In the first as well as in the second case, the ratio of the theoretical to the actual breaking load revealed a significant safety factor.

Tests on conventional structures using steel-reinforced polymer concrete columns and also foundation blocks and panels confirmed the basic theoretical assumptions adopted in the design. The State Institute for Designing Non-ferrous Metallurgical Plants on the basis of the results obtained produced working drawings of these structures, a large series of which were produced and set up at the Dzhezkazgan Mining and Metallurgical Combine. The introduction of reinforced polymer concrete structures for supporting baths helped to eliminate the use of expensive and deficient refractory-lined material, reduce the cost of construction, improve protection against electrocorrosion, reduce production labour by three times, increase the service life of structures by five times, and realise an annual economy of about 480,000 rubles.

The production of chemical-resistant polymer concrete tiles for lining the floors and storm-water drains has been perfected at the above works. To date the Dehezkazgan Mining and Metallurgical Combine has produced and laid over 5000 m^3 of reinforced polymer concrete structures.

The experience of using FA (furfural-acetone) polymer concrete for supporting mine shafts driven in frozen rocks is of great interest. At the Mikhailovsk iron ore combine, an experimental section of mine shaft was concreted to a height of 20 m with a wall thickness 500 mm. After each round of concreting, the heating of the polymer concrete laid was measured. After 20–30 min of laying to a thickness of 250 mm from the frozen wall, the polymer concrete temperature rose to 50–52°C, thus pointing to a normal polymerisation process. In all, 118 m^3 of polymer concrete were applied in the shaft. A month after work had stopped the polymer concrete support, exposed to freezing during this period, had a dense and uniform structure without spalling or crack formation. Control cube specimens showed a strength of 40–45 MPa, or roughly 2–2.5 times more than that of cement concrete specimens.

The polymer concrete mix was prepared in a mixer above ground and transported to the site of laying through hinged pipe lines. Easy workability was

Fig. 11.8. Polymer concrete column supporting bath platform. Column together with foundation block during tests.

ensured by using a large amount of furan resin (15–16% by weight). Although the high consumption of resin ensured good fluidity of the mix, it simultaneously increased the cost of the polymer concrete. Later, FA polymer concrete was used for concreting the tubbing space of many mine shafts of potash mines in Soligorsk and Bereznikakh. In these shafts, several thousands of cubic metres of polymer concrete were used.

In the Prokop'evskugol' Combine, a shop was set up for producing 1500 m³ per year of tubbings in coal mines using polymer concrete based on phenol-formaldehyde resins. However, utmost success was reported in developing and industrial adoption of baths including reinforced polymer concrete galvanic, etching and electrolytic baths.

Baths are of special importance in many branches of the industry, specially in non-ferrous metallurgy, chemical, petrochemical and several others. For example, the experience of producing containers based on polyester resins at the Kalushsk Chemical Combine is of great interest. Such containers are assembled from polymer concrete rings of diameter 1.6 to 2 m reinforced with prestressed steel reinforcement. This greatly simplifies the technology of producing large chemical-resistant containers.

It should be pointed out that the bath equipment operates under intense aggressive conditions as well as temperatures and loads. The problem is complicated by the fact that the baths are of extremely large dimensions. The diameter of thickeners ranges, for example, from 9 to 18 m with height exceeding 4 m Wetting towers of diameter up to 6 m rise to a height of 14 to 18 m or more.

The first lot of experimental-industrial polymer concrete baths reinforced with steel for electrolysis of copper were produced in 1960 at the Moscow Copper Smelting and Copper Electrolytic Works and Balkhashsk and Almalyksk Mining and Metallurgical Combines. In 1965, at the Moscow Copper Smelting and Copper Electrolytic Works, 30 single-piece electrolytic baths 2.46 m × 1.02 m × 1.2 m were produced using polymer concrete based on FAM (furfural-acetone) resin while the production of assembled polymer concrete baths commenced in 1972. These baths were lined with polyvinyl chloride plastic or polypropylene. More than 200 baths were produced at this works during 1972–1979 using steel reinforced polymer concrete based on FAM resin. The high corrosion resistance of these baths ensured their long service life without repairs.

The operational experience of single-piece steel-reinforced polymer concrete baths drew attention to their several vital drawbacks. In the process of production as well as during use, cracks were seen in them, which on widening, caused electrolyte leakage and consequent operational disadvantages.

An analysis carried out by the Reinforced Concrete Research Institute, State Institute for Designing Non-ferrous Metallurgical Plants, and Kazakh Non-ferrous Metallurgical Repair Trust established the following reasons for cracks in the polymer concrete electrolytic baths produced at the Moscow Works and Balkhashsk and Almalyksk Combines: the steel-reinforced polymer concrete baths were produced in the same moulds used for reinforced concrete baths; due to shrinkage processes and the generation of considerable compressive forces after moulding, it became difficult to take out the inner portion of the casing; when removing the case, microcracks formed which widened with use; further, high stresses arose in the walls, bottom and corners of the bath due to high rigidity of the casing and steel reinforcement; the structural rigidity of the bath as a whole was not adequate and no methods were available for designing such structures.

A special casing with a flexible core and collapsible sides was designed to eliminate the above sources of crack formation and a design method worked out. The bath structure was strengthened and the inner corners of the bath chamfered to a radius of 125 mm to reduce stresses.

The use of steel reinforcement for electrolytic baths is also responsible for the possible generation of cracks. In the case of diffusion penetration, electrolyte over the course of time comes into contact with the steel reinforcement and, in the presence of an electrical potential, the non-ferrous metal (copper and zinc) deposits on it disrupting the entry of polymer concrete which leads to consequent crack formation.

Glass-plastic reinforcement possesses high corrosion resistance in addition to being a good dielectric. The reinforcement of electrolytic baths with prestressed glass-plastic improves the crack resistance of the structure in the course of production and use. The glass-plastic reinforcement was prestressed to 630 MPa, i.e., 45% of its short-term strength.

The adoption of these technical solutions helped in designing reliable structures of single-piece electrolytic baths and in organising their commercial production at several plants which to date have produced more than 2000 of them.

The introduction of reinforced polymer concrete baths and structures made possible the elimination of various chemical-resistant protective refractories and economise 1500 tonnes of lead, a large amount of graphite blocks and other deficient materials and, no less important, improved the quality of the non-ferrous metals produced.

The total volume of reinforced polymer concrete structures introduced in the non-ferrous metallurgical works exceeds $30,000 \, m^3$ with a consequent annual economy of about 10 million roubles.

The experience of developing and producing polymer concrete structures at the non-ferrous metallurgical works was utilised by other ministries and departments. For example, the Dnepropetrovsk Engineering and Design Institute introduced large-sized etching baths of length about 12 m and weighing up to 80 tonnes at the Dnepropetrovsk pipe Rolling Works.

When producing such containers, the inner portion of the casing was provided with flexible compensators for shrinkage strains. The number and thickness of elastic liners were determined theoretically, taking into consideration the linear shrinkage of the polymer concrete. Moreover, plasticisers were added to the polymer concrete composition to reduce shrinkage strains.

To prevent the formation of cracks between the foundation and bottom of the etching bath, a partition consisting of two layers of water-proofing material with graphite powder between them was provided. Graphite powder ensures free

Fig. 11.9. Machine frame made of polymer concrete.

sliding of the bath bottom along the base to allow for both shrinkage strains and thermal strains during its use.

In the etching division of the plant, all the etching baths made of a metal body with highly efficient lining of acid-resistant refractory bricks based on andesite plaster were replaced by polymer concrete baths. The exploitation of polymer concrete etching baths containing 20% sulphuric acid at 45–75°C showed that polymer concrete of the selected composition possesses high chemical resistance, good abrasion resistance and impact strength. Process tanks for hot (up to 70°C) sulphuric acid of up to 40% concentration were also introduced in this plants.

Table 11.1 shows the most popular range of polymer concrete structural members based on various resins used in the Soviet Union. Table 11.2 shows the range proposed for manufacture up to 1995.

The designs of industrial buildings and installations using polymer concretes are shown in Fig. 11.10.

268

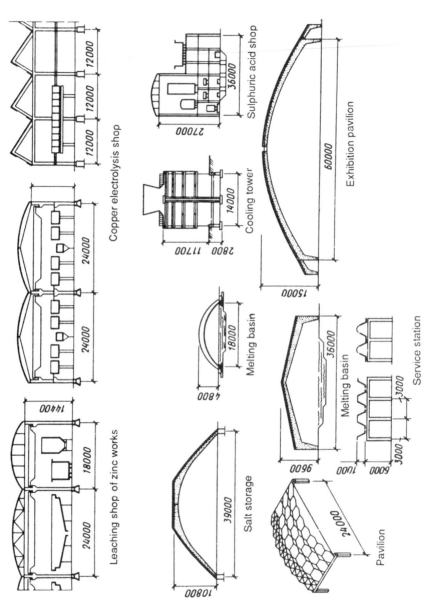

Copper electrolysis shop

Leaching shop of zinc works

14400

18000

24000

Sulphuric acid shop

27000

36000

Cooling tower

11700 2800

14000

Melting basin

4800

18000

Salt storage

10800

39000

Exhibition pavilion

15000

60000

Melting basin

9600

36000

Service station

6000

1000

3000

1000

3000

Pavilion

24000

Fig. 11.10. Principal designs of industrial buildings and installations using polymer concrete structures.

The American company Bondite produces decorative tiles and wall panels made of polymer concrete based on phenol resins and organic fillers in the form of sawdust, nut shells, rice husk, etc. Polymer concretes based on organic fillers have low density at a compressive strength of up to 40 MPa. The same

Table 11.1. Range of structural components made of polymer concretes used in the construction of industrial buildings

Item	Sketch of structure
Frame elements	
Building frame foundation	
Foundation for equipment	
Frame columns for multistoreyed building	
Columns for built-in stacks and platforms under equipment	

(Contd.)

Table 11.1. (*Contd.*)

Item	Sketch of structure

Tiles for covering (flat, ribbed,
   double-T type, closely ribbed, etc.)

Beams and internal elements (rectangular,
   T-shaped, double T-shaped, etc.)

KZhS Camel-back shaped units

Supports

Transmission line members

Bunkers

Elements of process equipment

(*Contd.*)

Table 11.1. (*Contd.*)

Item	Sketch of structure

Electrolytic baths (copper, zinc, nickel, etching, etc.)

Lining (blocks, tiles, etc.)

Cantilevers, grids made of sleepers, boards, etc. under process pipe-lines

Members for cleaning equipment and process water storage

Pipes (collectors) for acid discharges

Channels (troughs), wells

(*Contd.*)

272

Table 11.1. (*Contd.*)

Item	Sketch of structure
Settlers	

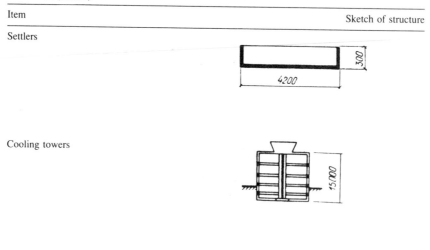

Cooling towers	

company has developed the technology and specialised equipment for producing polymer concrete drainage pipes, radio-impervious panels, blocks with carbon-containing fillers and polymer concrete based on MMA (methyl methacrylate resin) for military purposes, including special buildings and operational repair of airport runways.

In Japan, three stations for measuring the earth's magnetism were built using polymer concrete panels reinforced with glass-plastic.

11.9. Use of Polymer Concretes in Hydraulic Engineering and Reclamation

It is known that most rivers of central Asia, Trans-Caucasus and some other regions of the Soviet Union are fed by waters consequent to thawing of high-altitude snows, glaciers and precipitations in their catchment areas. These factors make for extremely uneven distribution of water flow in the annual cycle. In the autumn-winter period the water flow decreases sharply and in the high-water season rises abruptly. In this period, large amounts of bed sediments ranging from fine sand to large boulders weighing several hundred kilograms are carried into the montane and foremontane sections of rivers.

In the high-water season, intense wear of hydraulic spillway installations is noticed. This wear depends not only on the size and saturation of the current with sediment, but also on the rate of flow, degree of rolling and mineral composition of the drifting material. Practical observations have shown that suspended matter of 0.05–0.5 mm in size contributes to the wear of equipment surfaces to a very insignificant extent; but at flow rates of 25–30 m/sec or more, independent of

Table 11.2. Range of polymer concrete structural components for use in constructing industrial and social buildings

Item	Sketch of structure

Frame elements

Pile foundation (tubular piles)

Γ-shaped frames

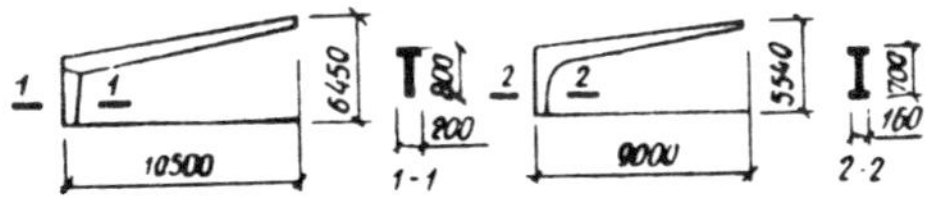

Curvilinear frames

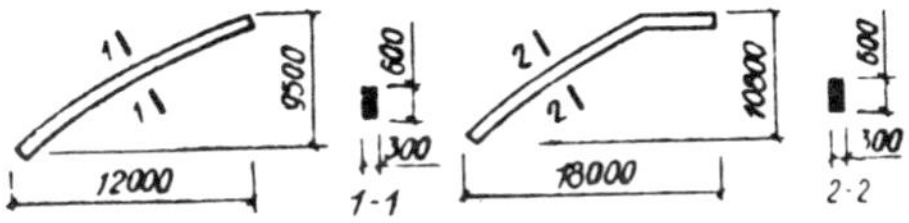

Truss

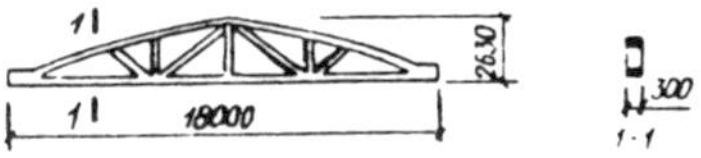

Domes

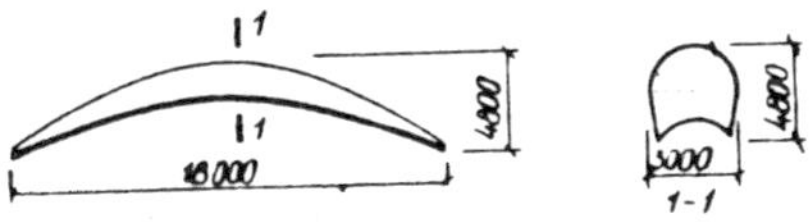

Folded plates

(Contd.)

Table 11.2. (*Contd.*)

Item	Sketch of structure
Grids	

Frame columns for single-storey
 buildings

Elements of process equipment

Ventilation ducts

Thickeners

the content of suspended matter, surface damage is possible due to the effect of cavitation also.

Under such operating conditions of hydraulic equipment, some types of polymer concretes compete successfully with such conventional materials as natural rocks, cast iron, etc.

In 1961–1964, some river rapids of spillways of dams in central Asia were lined with monolithic polymer concrete based on furan resins to a thickness of 100–150 mm. The total covered area on five dams was about 900 m². Over 25 years of operational experience these dams have shown the high wear resistance, durability and economic advantage of using polymer concrete coatings.

Later, for protecting dams, not only monolithic polymer concrete based on

FAM and FAED resins, but also precast polymer concrete tiles based on these same resins have been produced under plant conditions.

Furan polymer concrete was used as an anti-cavitational lining for the first time in 1960 in the Volga Hydroelectric Power Station where cavitational damages were noticed in the baffle blocks of spillway aprons and also on the bottom spillways. Later, anti-cavitation coatings of polymer pastes and plasters based on epoxide compositions were used at Bratsk, Nureksk, and Charvaksk Hydroelectric Power Stations. After 15 years of use, the coatings continued to be in a satisfactory condition.

The engineering and research centre of water and power resources (Denver, Colorado), Brookhaven National Laboratory (Apton, New York), waterways experimental station of the Army Engineering Corps of the USA (Vicksburg, Mississipi) and several large companies have been carrying out major studies on developing compositions, production technology and the use of polymer concretes, polymer plasters and pastes as protective and wear-resistant materials for various hydraulic installations.

The use of polymer protective compositions based on epoxide resins at more than ten hydraulic engineering installations in the regions of Vicksburg (Mississipi), Kansas City (Missouri), Pittsburg (Pennsylvania), Los Angeles (California), Little Rock (Arkansas) and others provided wholly satisfactory results. Similar studies are being carried out in France and other countries.

The Japanese firm Taisei Corporation has developed the technology for guniting polymer plasters based on epoxide resins for repairing concrete structures and also the technology for repairing worn-out concrete tunnels used for carrying sewage and irrigation waters (Zenitaka Corporation). Similar work is being carried out in many other developed countries.

Among chemically stable reinforced polymer concrete structures, pipes occupy an important position. The first of experimental-industrial sewer pipes or rings made of polymer concrete based on FA resin (furfural-acetone) were produced at the Skuratovsk works in 1961. With an internal diameter of 1680 mm, wall thickness of 150 mm, and length of 1200 mm, the pipes weighed 2270 kg including 252 kg of steel reinforcement. The latter consisted of two cylindrical grids of hot-rolled wire of diameter 5 mm with a grid size of 100 mm × 100 mm and joined with cross-rods. Polymer concrete sewer pipes were made in a special unit by vertical vibration moulding. The first crack in the pipes tested appeared at a load double that on similar pipes made of reinforced concrete grade V40. The breaking load of polymer concrete pipes went up to 250–300 kN/m.

The first experimental sewer of length 38 m was laid under the Novo-Ryazansk highway by adopting an advancing technique and using polymer concrete pipes. Using an advancing device (in 1962), a similar sewage section of length 72 m was laid in the region of Tsaritsyno-Vidnoe. A month after laying the test section, tests for leakage were conducted throughout the sewer length. The sewer, consisting of a single layer of polymer concrete pipes, withstood the tests very satisfactorily. Its hermetic sealing was twice better than that of reinforced concrete sewer with a protective lining.

Work has been carried out in several organisations on selecting and studying the compositions and production technology of drainage and water conduits made of polymer concrete based on carbamide resins. Technical specification TU 33-112-79 for joined polymer concrete pipes, approved by the USSR Ministry of Water Management, covers pipes of diameter 300–1000 mm and length up to 5 m.

The polymer concrete mix was made in concrete mixers and polymer concrete pipes were produced by the centrifugal method in standard centrifuges designed for producing reinforced concrete pipes. Moulding was done in standard metallic detachable moulds consisting of two half moulds and connecting rings.

An essential feature of moulding polymer concrete pipes based on UKS resin by the centrifugal method is that, during compaction, free water is partially exuded and there is no liberation of the binder on the inner surface. The partial elimination of water improves the structure of polymer concrete and its physical and mechanical properties while the absence of binder liberation on the inner surface of the pipes makes blowing with hot air unnecessary.

Polymer concrete pipes based on UKS resins are designed for constructing irrigation systems. Such pipes can be used in aggressive media with pH 3 to 10 and as conduits with pressure up to 0.2 MPa at a depth of up to 5 m Pipe sections of diameter 600 mm and total length 2400 m laid in 1970 revealed no damage after eight years of use.

The large-scale production of polymer concrete pipes based on polyester resins was organised in Japan by Hokusan Rezikon, Kurimato Airon Works, etc. Polymer concrete pipes are produced by two methods: centrifuging and extrusion. Pipes made by extrusion have a cross-section of polymer plaster in multiple layers and are reinforced with fibreglass. Kubota company produces polymer concrete pipes of diameter about 5.5 m with an annual capacity of 30,000 tonnes.

The American company Amoko Techeit (California) has constructed a large plant for the commercial production of polymer concrete pipes of diameter 300 to 3000 mm with a wall thickness of 8–15 mm. The raw materials used are (% by weight): polyester resin with appropriate hardeners 26–35; quartz sand of size fraction 0.2–0.5 mm, 40–45; fibreglass reinforcement material (glass roving, fibreglass matting, etc.) 30–35.

Depending on the diameter and pressure rating, the cost of pipes varies roughly from $30 to $100 per metre. On average, the plant production capacity per shift is 100 m in terms of 1500 mm diameter and 6100 mm long pipes. According to the plant data, the durability of such pipes is at least 50 years. The product competes successfully with steel, asbestos cement and reinforced concrete pipes. The pipes produced by this company are used in high-pressure mainlines for discharging aggressive products in cleaning installations and for pumping brine and also as casing pipes in drilling wells, for pumping oil, etc.

The production of polymer concrete pipes of large diameter by radial pressing and centrifuging has been perfected by several firms in the FRG.

Much research is being carried out in Czechoslovakia on polymer concretes based on furan resins and furyl alcohol. Among the more interesting studies are the production of pipe lines for handling aggressive industrial discharges using polymer concrete pipes of diameter 1350 mm, length 3000 mm and wall thickness 40 mm. Pipes in two layers to diameter 1800 mm and length 2000 mm are also produced using cement concrete with an internal polymer concrete layer of 20–30 mm by successive centrifuging or vertical vibration moulding.

Polymer concrete was used in large quantities in laying the new main sewer line intersecting Prague from south to north. The total length of this sewage line is more than 11 km. The inner diameter of the line at different sections ranges from 2000–3600 mm. Some original innovations were made in the design: in a specialised plant, segments of polymer concrete based on furan resins were produced by vibration pressing in lengths of 1490 mm and width 900–1000 mm with the required curvature. The weight of each segment is 60–70 kg.

These segments were assembled in a specially designed metal casing of length up to 3000 mm with its outer diameter corresponding to the inner diameter of a given section of the line. The segments were drawn by a wire and the joints between the segments closed with temporary rubber linings. Later, the casing together with the segments was placed in position by a crane when laying in an open trench or advanced using special trolleys when laying in a tunnel. After mounting two or three sections, the line was joined with cement concrete.

After the concrete acquired adequate strength, the casing was withdrawn from the line on a truck. Later, the rubber linings were removed and the joints between the segments filled with polymer plaster based on epoxide resins.

Compared to the original programme, which provided for the use of a protective lining of acid-resistant bricks on the inside surface, the present design reduced the labour of laying to one-third in tunnels and to one-sixth in open trenches. The labour economy was 70%.

11.10. Polymer Concretes with Special Properties

By appropriate selection of the compositions and technological processes, it is possible to produce polymer concretes possessing special and, in many cases, unique properties. In radio and electrical engineering industries, and in nuclear power, machine and tool fabrication industries, polymer concretes are finding ever increasing use because of their easy workability, comparative cheapness and, when needed, with high dielectric or electrical conducting characteristics, high damping properties and resistance to various radiations.

Polymer concretes with high dielectric characteristics

It is known that the dielectric characteristics of most polymer materials are the result of their molecular and supramolecular structures. The dielectric properties of some types of polymer materials are shown in Table 11.3.

Table 11.3. Dielectric characteristics of some types of polymer materials

Material	Dielectric constant	Tangent of dielectric phase angle	Specific surface electrical resistance, ohms	Specific volume electrical resistance, ohm-cm
Polystyrene emulsion	2.6–3	0.001–0.002	10^{14}–10^{15}	10^{14}–10^{15}
Common polystyrene	2.6	0.00045	10^{15}	10^{16}–10^{18}
Polyvinyl chloride	3.5	0.03–0.05	–	10^{14}–10^{16}
Methyl methacrylate	3.2–3.6	0.02–0.06	–	10^{12}
Polyethylene	2.2–2.3	0.0002–0.0004	10^{17}	10^{17}

Most of the dielectrics developed on the basis of polymers are mainly used in the form of filled thermoplastics or pastes using thermosetting oligomers for producing comparatively small components and products. Further, the maximum degree of filling such compositions does not exceed 40–50% in most cases.

A new and extremely promising direction is the development of special types of polymer concretes conforming to the modern technical requirements of dielectric properties. The use of appropriate oligomers, hardeners and fillers with optimum size composition makes for the production of polymer concretes with high physical, mechanical and dielectric properties as well as chemical resistance.

In this case, polymers are used most effectively since the extent of filling rises up to 85–90% and the material acquires structural properties and can be used for producing large-sized products and structures.

It should be borne in mind that a high degree of filling with mineral fillers and aggregates greatly modifies the nature of dielectric losses of polymer concretes. As already pointed out, the addition of mineral fillers and aggregates to the system results in the formation of typical supramolecular structures of the polymer binder. Such compositions exhibit electrical losses associated with the heterogeneity of polymer concrete mixes and structural losses which represent a consequence of some polarisation of binder molecules close to the surface of fillers and aggregates. Therefore, one of the most important problems associated with the selection of polymer concrete compositions with high dielectric parameters is to find ways that reduce the influence of the above factors on the dielectric, physical and mechanical properties of such highly filled compositions.

pressure mainlines for discharging aggressive products in cleaning installations and for pumping brine and also as casing pipes in drilling wells, for pumping oil, etc.

The production of polymer concrete pipes of large diameter by radial pressing and centrifuging has been perfected by several firms in the FRG.

Much research is being carried out in Czechoslovakia on polymer concretes based on furan resins and furyl alcohol. Among the more interesting studies are the production of pipe lines for handling aggressive industrial discharges using polymer concrete pipes of diameter 1350 mm, length 3000 mm and wall thickness 40 mm. Pipes in two layers to diameter 1800 mm and length 2000 mm are also produced using cement concrete with an internal polymer concrete layer of 20–30 mm by successive centrifuging or vertical vibration moulding.

Polymer concrete was used in large quantities in laying the new main sewer line intersecting Prague from south to north. The total length of this sewage line is more than 11 km. The inner diameter of the line at different sections ranges from 2000–3600 mm. Some original innovations were made in the design: in a specialised plant, segments of polymer concrete based on furan resins were produced by vibration pressing in lengths of 1490 mm and width 900–1000 mm with the required curvature. The weight of each segment is 60–70 kg.

These segments were assembled in a specially designed metal casing of length up to 3000 mm with its outer diameter corresponding to the inner diameter of a given section of the line. The segments were drawn by a wire and the joints between the segments closed with temporary rubber linings. Later, the casing together with the segments was placed in position by a crane when laying in an open trench or advanced using special trolleys when laying in a tunnel. After mounting two or three sections, the line was joined with cement concrete.

After the concrete acquired adequate strength, the casing was withdrawn from the line on a truck. Later, the rubber linings were removed and the joints between the segments filled with polymer plaster based on epoxide resins.

Compared to the original programme, which provided for the use of a protective lining of acid-resistant bricks on the inside surface, the present design reduced the labour of laying to one-third in tunnels and to one-sixth in open trenches. The labour economy was 70%.

11.10. Polymer Concretes with Special Properties

By appropriate selection of the compositions and technological processes, it is possible to produce polymer concretes possessing special and, in many cases, unique properties. In radio and electrical engineering industries, and in nuclear power, machine and tool fabrication industries, polymer concretes are finding ever increasing use because of their easy workability, comparative cheapness and, when needed, with high dielectric or electrical conducting characteristics, high damping properties and resistance to various radiations.

Polymer concretes with high dielectric characteristics

It is known that the dielectric characteristics of most polymer materials are the result of their molecular and supramolecular structures. The dielectric properties of some types of polymer materials are shown in Table 11.3.

Table 11.3. Dielectric characteristics of some types of polymer materials

Material	Dielectric constant	Tangent of dielectric phase angle	Specific surface electrical resistance, ohms	Specific volume electrical resistance, ohm-cm
Polystyrene emulsion	2.6–3	0.001–0.002	$10^{14}-10^{15}$	$10^{14}-10^{15}$
Common polystyrene	2.6	0.00045	10^{15}	$10^{16}-10^{18}$
Polyvinyl chloride	3.5	0.03–0.05	–	$10^{14}-10^{16}$
Methyl methacrylate	3.2–3.6	0.02–0.06	–	10^{12}
Polyethylene	2.2–2.3	0.0002–0.0004	10^{17}	10^{17}

Most of the dielectrics developed on the basis of polymers are mainly used in the form of filled thermoplastics or pastes using thermosetting oligomers for producing comparatively small components and products. Further, the maximum degree of filling such compositions does not exceed 40–50% in most cases.

A new and extremely promising direction is the development of special types of polymer concretes conforming to the modern technical requirements of dielectric properties. The use of appropriate oligomers, hardeners and fillers with optimum size composition makes for the production of polymer concretes with high physical, mechanical and dielectric properties as well as chemical resistance.

In this case, polymers are used most effectively since the extent of filling rises up to 85–90% and the material acquires structural properties and can be used for producing large-sized products and structures.

It should be borne in mind that a high degree of filling with mineral fillers and aggregates greatly modifies the nature of dielectric losses of polymer concretes. As already pointed out, the addition of mineral fillers and aggregates to the system results in the formation of typical supramolecular structures of the polymer binder. Such compositions exhibit electrical losses associated with the heterogeneity of polymer concrete mixes and structural losses which represent a consequence of some polarisation of binder molecules close to the surface of fillers and aggregates. Therefore, one of the most important problems associated with the selection of polymer concrete compositions with high dielectric parameters is to find ways that reduce the influence of the above factors on the dielectric, physical and mechanical properties of such highly filled compositions.

In turn, the rational application of polymer concretes as efficient dielectrics involves not only the development of suitable compositions and obtaining reliable data on their dielectric parameters, but also calls for a thorough study of the variation patterns of their properties in relation to the production technology, thermal treatment regime and operational conditions.

It is known that polarisation (movements) of charged particles (atoms, molecules, and ions) occurs in dielectrics and semi-conductors placed in an electrical field. In dielectrics, although this movement is insignificant, it nonetheless constitutes an important characteristic. Polarisation generates an additional electrical field in the material whose force lines fall opposite the lines of the external electrical field and weaken it. Moreover, as a result of the friction of particles amongst themselves, polarisation is accompanied by losses of field energy which cause additional losses resulting in the molecules being incapacitated from orienting beyond the external field and polarisation stopping short of it. The movement of the orbit and the rotation angle of polar molecules are greater, the higher the electrical field voltage E. The number of movements of the orbit and the dipole rotations in a unit of time are directly related to the frequency of electrical field fluctuations f. Consequently, the main parameters of the electrical field are the voltage E (kV/cm) and frequency f (MHz). At the same time, in a given electrical field with known frequency, different materials undergo a different degree of loss directly depending on the properties of the material itself. The basic electrical parameters determining the properties of the dielectric material are: dielectric constant ϵ, tangent of dielectric phase angle $tg\,\delta$ and electrical resistance R. The lower the values of ϵ and $tg\,\delta$ and the higher the value of R, the better the dielectric characteristics of a given material. Hence, the problem of developing new types of dielectrics lies in formulating materials in which the values of dielectric constant ϵ and tangent of dielectric phase angle $tg\,\delta$ are as minimal as possible while the physical and mechanical characteristics are maintained at a high level.

In the course of developing polymer concrete compositions with high dielectric parameters at the Reinforced Concrete Research Institute, carbamide, phenol-formaldehyde, polyester, furan, epoxide and other resins were tested as polymer binders in combination with various mineral fillers. The dielectric parameters of raw materials are shown in Table 11.4.

Preliminary investigations of these polymer concretes showed that their dielectric parameters are influenced not only by the type of polymer binder and fillers selected and their percentage ratios, but also by the type and amount of hardener, methods of preparation, moulding and hardening of the polymer concrete mixes.

From among the compositions developed, the best results of dielectric parameters were obtained for light polymer concretes based on furan resin FAM and claydite aggregate. Among the deficiencies of these polymer concretes are the comparatively low strength and sharp impairment of dielectric parameters

Table 11.4. Dielectric parameters of polymer binders and mineral fillers

Material	Dielectric constant ϵ	Tangent of dielectric phase angle tg δ	Electric resistance R, ohms
Furan resin*	2.5–3	0.05–0.06	2–3 ($10^8 - 10^9$)
Epoxide resin ED-16	3.5–6	0.01–0.03	$10^{12} - 10^{13}$
Furan-epoxide resin (40–60)	2.15–2.3	0.05–0.06	5–6 ($10^8 - 10^{11}$)
Granite rubble**	1.9–2	0.05–0.05	–
Quartz sand	1.9–2.2	0.025–0.03	–
Claydite	1.5–1.8	0.03–0.05	–
Andesite flour	3.5–3.8	0.045–0.05	–
Phenol resin	4.5–5	0.015–0.03	10^9–10^{12}
Polyester resin PN-1	2.8–5.2	0.011–0.035	10^{11}–10^{14}

*After thermal treatment of oligomers at 80°C for 24 h.
**Fillers were dried to constant moisture.

after 30 days of retention in water (Table 11.5). In the case of heavy polymer concretes, good results were obtained when using furan resin FAM and furan-epoxide compound FAED. While possessing extremely high strength indices in the dry state, their dielectric parameters are roughly identical. The dielectric characteristics of polymer concretes based on FAED binder underwent almost no change after 30 days of water saturation while the value of tg δ of polymer concretes based on furan binder rose fivefold.

Table 11.5. Main characteristics of light and heavy polymer concretes

Index	Light FAM	Heavy FAM	Heavy FAED
Average density, kg/m³	1600–1800	2200–2300	2200–2300
Compressive strength, MPa	25–30	80	120
Modulus of elasticity, MPa	–	28×10^3	30×10^3
Maximum elasticity	–	24	42
Poisson's ratio	–	0.2–0.22	0.29
Dielectric constant ϵ	3.1	4	4
Tangent of dielectric phase angle tg δ	0.044	0.05	0.045
-do- after 30 days of retention in water	0.41	0.26	0.05

Based on the nature of variation of dielectric parameters of these three types of polymer concretes in relation to the duration of retention in water, it can be concluded that light polymer concretes based on FAM can be recommended only for dry conditions of use or they should be well protected by water-resistant, impermeable coatings. Heavy polymer concretes based on FAM are less sensitive to water but, on prolonged use in humid conditions or when water saturation is likely, they, too, need good protective coatings. Polymer concretes based on FAED have maximum strength characteristics and stable dielectric parameters and require no protective coatings but their high cost is an inhibiting factor.

The impairment of the properties of polymer concretes on wetting is reversible. On driving out the moisture, the dielectric parameters are restored.

The compositions of polymer concretes developed with high dielectric parameters meet the contemporary requirements, widen the range of materials used in the electrical and radio engineering industries, and help to resolve many structural and operational problems by very simple means.

For example, the American company Lindsay Industries (California) has been carrying out work from 1977 on producing insulators for high-voltage electrical transmission lines and elements of dielectric protective facings for electrical communications for underground power facilities using polymer concretes based on polyester and other resins.

Comparative tests with porcelain and polymer concrete insulators showed that electrical breakdown of porcelain insulators occurred at 235 kV and of polymer concretes at 250 kV.

Comparative data on the properties of porcelain and polymer concrete (by the method of M. Gunasekaran) are shown in Table 11.6.

There is yet one more advantage of polymer concrete, i.e., the possibility of introducing metallic elements into the mass during casting. Thus, when producing tower insulators and insulators for supports, metal caps can be introduced into the mass directly during production. Moreover, the inserted metal serves as a voltage adapter and ensures uniform voltage distribution on the insulator surface. This property improves the polymer concrete characteristics by 20% under conditions of variable (60 Hz) as well as direct voltages.

Polymer concrete insulators have been used for several years in the USA. For example, many stations of the Pennsylvania Power and Light Company use insulators of 25 cm height (12 kV) as insulator-regulators (Fig. 11.11).

In Houston, insulators of height 115 cm were designed. Following tests on prototypes, they began being used in 1984 in Pennsylvania in the external installations of power stations. They constitute the largest single-piece insulators in the world. These insulators worked initially at voltages of up to 69 kV and later up to 138 kV. The insulators were designed taking into consideration the prevailing specifications of the American National Standards Institute for porcelain insulators of height 115 cm.

In formulating the optimum composition, innumerable recipes were tested using extremely divergent fillers and resins. The selection of a specific recipe for producing polymer mixes was based on the technological workability of the mix under vacuum and vibration conditions, mechanical stability of the mix over a prolonged period under conditions of atmospheric variations and cost of the material.

282

Table 11.6. Physical and technical properties of porcelain and polymer concrete

Index	Porcelain	Polymer concrete
Processing temperature, °C	1000 and above	Room temperature, up to 120°C
Processing duration	Many hours and days	A few minutes to 4 h
Dielectric constant	5.4–7.5	4.1–5.5
Dielectric loss factor	0.009–0.025	0.011–0.030
Dielectric breakdown strength, kV/cm	110–600	500–900
Volume resistance, ohms/cm	10^{13}–10^{15}	10^{15}
Bending strength, MPa	50–120	50–195
Compressive strength, MPa	150 (1000 on small specimens)	180–250
Metal joint	Not permitted during firing	Permitted during casting
Reinforcement	Possible only by external device (mechanical)	Reinforcement possible with various fibres or rods with dielectric properties
Building up by layers	Impossible because of high firing temperatures	Possible; resistances can be laid inside the insulator or various mixes can be formulated or reactive resistances can be introduced into the mix
Increasing blast resistance	Impossible	Possible by introducing fibres or grids
Investment	High	Low

Insulators of this type were subjected to severe high-voltage tests in the IREK high-voltage testing laboratory at Montreal (Canada) and mechanical tests for compression, breaking, torsion and cantilever strength. These tests confirmed the high efficiency of polymer concrete.

Fig. 11.12 depicts a superinsulator assembly made of polymer concrete rings each weighing one tonne.

A series of polymer concrete insulator bases, a large number of distribution boxes and other products are produced in the FRG; the USSR produces a series of cross-pieces for 10 kV electrical transmission supports and cable boxes using polymer concrete based on FAED (Fig. 11.13).

In Japan, polymer concretes based on wide-ranging polyester resins developed by Nippon Telegraph and Telephone are used for making manholes of telegraph cables. At present, four Japanese firms produce about 20,000 tonnes a year of polymer concrete manholes, or roughly 30% of their total production.

Fig. 11.11. Testing insulators made of polymer plaster for electrical breakdown.

Conducting polymer concretes

The development of modern science and technology calls for a steady expansion of the manufacturing of the latest means of automation, computer technology, and radio-electronic devices and equipment used in extremely diverse branches of science and technology. As a result, the land space is becoming increasingly saturated with radiations from radio engineering, electrical and other installations. Literature reveals that the number of radiation sources is doubling while the electromagnetic energy radiated is increasing ten times every decade. Electromagnetic radiations can rationally be regarded as one type of environmental contamination. They adversely affect the health of people and interfere in the work of various radio-electronic installations. Thus, the working of the latter without impairing the quality indices is becoming increasingly complex while environmental protection is becoming an increasingly urgent problem.

284

Fig. 11.12. Superinsulator assembled from polymer concrete rings each weighing one tonne.

Fig. 11.13. Coupling boxes made of polymer concrete based on FAED for electrical cables.

Fairly reliable means of protection from adverse electromagnetic radiations on the serving personnel and equipment have now been developed and perfected. The most rational methods of such protection are based on engineering and technology directly aimed at reducing the intensity of electromagnetic radiations to the permissible level. Screening and protective filters are among such measures. Protective filters reduce the interferences penetrating into the device through the feeder mains. Screening effectively protects the electronic equipment from external interferences and simultaneously represents a reliable method of controlling the natural radiations in the environment. Screening of buildings housing electronic equipment is effected by lining the walls with special conducting materials.

Various metals with good screening properties are used as screening materials. These ensure good screening in a wide range of electromagnetic field intensities. However, the use of metallic sheathings for screening the buildings suffers from certain deficiencies, such as comparatively high cost and limited service life since many metals undergo corrosion, reduce the comfort inside the buildings, etc.

Therefore, work is being carried out in many countries on replacing metal screens with other conducting materials including those based on polymers. The problem involves producing materials with properties as close as possible to those of the main screen, i.e., in the degree of weakening of the electromagnetic field energy penetrating the screen. The degree of weakening or the efficiency (Ef) of screening is represented by the ratio of the electric field voltage E_1 or magnetic N_1 component at a given point without the screen to the field voltage at the same point when using a screen, E_2 or N_2:

$$Ef = E_1/E_2 \quad \text{or} \quad Ef = N_1/N_2.$$

The efficiency of screening depends directly on the electrical conductivity of the material used. Therefore, only such non-metallic materials which have a resistance not exceeding 10 ohms can compete with metallic screens. The development of conducting materials based on polymers is a very complex task compared to the development of dielectrics based on polymers.

However, the comparative simplicity of processing and application of protective coatings has drawn much attention to the use of polymers for developing conducting materials in form of enamels, pastes, mastics and adhesives. For developing such conductors, materials based on rubbers and epoxide resins were studied more fully than compositions based on polyesters, phenol-formaldehyde, furan, and polyurethane resins, polyvinyl acetate, polyvinyl chloride, polystyrene and other polymers.

It is known that most polymers are good dielectrics, i.e., have a very high electrical resistance in the range 10^8 to 10^{16}. Therefore, in developing conducting materials based on polymers, many difficulties have to be overcome, primarily the selection of fillers with high conducting properties. It is quite natural to assume that powders of various metals represent such fillers. Several works of western specialists provide data on the use of powders of silver, copper, nickel, tin, aluminium, iron, ferromagnetic alloys, etc. as conducting materials. However, subsequent investigations revealed that the addition of most of the aforementioned metallic powders to the polymer composition did not provide the expected results. This is explained by the comparatively rapid formation of oxide films on the surface of finely dispersed particles of several metals as a result of which their resistance increases sharply. Very good results were obtained on adding silver powder with a flaky form of particles or nickel powder to polymer compositions but these powders are very expensive as well as scarce and hence can hardly be used on a large scale. Graphite or powdered coke, soot, carbonised viscose fibre type 'Uglen', etc. were also tried as fillers for conducting compositions.

The nature of the binder likewise exerts a significant influence on the electrical as well as physical, mechanical and operational properties of conducting compositions. The polymer binder exerts a direct and dominant influence on the electrical resistance of the polymer composition only when the extent of filling with the conducting fillers is comparatively low, i.e., as long as an electrically conducting structure is not formed in the system. On further increasing the filler content, the factor most influencing the electrical conductivity of the composition is not the electrical resistance of the binder, but the packing density of the filler since the greater the number of grains, the greater the number of contacts. It is these contacts that determine the electrical conductivity of the composition as a whole.

The physical and mechanical properties of conducting materials depend on the ability of the polymer binder to wet well the particles of the selected conducting fillers to form fairly high adhesion bonds. In turn, the filler particles should disperse well in the selected polymer binder. When the compatibility between the binder and the filler is poor, the particles of the latter aggregate in the mix render the formation of continuous conducting structures difficult and impair the physical and mechanical properties of the product.

To improve compatibility with the conducting fillers and their uniform distribution in the mix, surface-active agents are generally added to the composition while solvents or diluents of polymer binder are added to improve the contacts between the filler grains. The technology of production also greatly affects the properties of the conducting materials.

An analysis of the published literature, including patents, showed that in spite of the large number of researches carried out, conducting materials based on polymers are used mainly for producing heater elements, for tapping static electricity, and for adhesives, pastes, and paint and varnish coatings.

An extremely interesting conducting material based on mineral wool and polyacrylamide filled with soot was recently developed (Table 11.7).

Table 11.7. Main properties of electrically conducting material based on mineral wool and polymer binder

Index	Production technology	
	paste application	casting
Average density, kg/m^3	450–500	450–500
Maximum bending strength, MPa	0.8–1.2	1.7–2.5
Damping energy of electromagnetic waves at 3000 MHz, dB/cm	24–28	26–30
Resistance, ohms	10	10
Water absorption, %	2–3	2–3

The material has a resistance of 10 ohms. The performance of a screen made from this material is similar to that of a metallic screen. Damping of the energy of electromagnetic waves is mainly due to its reflection from the screen surface; only an insignificant part of the energy is dispersed in the form of heat within the screen itself.

An essentially new direction is the development of conducting materials based on conducting polymer concretes. The importance of this direction is the many advantages of these materials over existing ones. The electrically conducting polymer concretes are comparatively easily produced in the form of complex products and structures. Their high strength characteristics make it possible to design load-bearing and self-supporting structures. They possess high corrosion resistance and are less scarce compared to non-ferrous metals and can be used as screening materials not only against electromagnetic waves, but also various radiations.

Based on the data available in the published literature and patents for producing conducting polymer concretes, the same resins used for producing dielectrics were tested. In these tests, coke and graphite fractions (1.0–5 mm) were used as aggregates; graphite powder with specific surface 3000 cm^2/g; soot with specific surface 20 m^2/g and metallic powder with specific surface 2000–3000 cm^2/g were used as fillers. The specific body resistance of the fillers was as follows, ohm-cm: metallic powder 10^{-2} to 10^{-3}; graphite 10^{-2}, soot 10 to 10^{-1}; and coke 10 to 10^{-1}.

Preliminary investigations of electrically conducting polymer concretes based on various oligomers showed that carbamide resins can be used to produce comparatively high conducting characteristics (specific body conductivity 8–9 siemens-cm). However, the ultimate compressive strength of such polymer concretes is quite low (6–6.5 MPa). Considerably better results were obtained in the case of polymer concretes based on furan, epoxide, and phenol-formaldehyde resins (Table 11.8). Carbon-bearing materials together with metallic powders were used as fillers and aggregates.

Table 11.8. Main characteristics of electrically conducting polymer concretes

Index	Polymer concretes based on binder		
	FAM (furan)	ED-20 (epoxide)	SFZh (phenol-formaldehyde)
Average density, kg/m^3	1280	1130	1350
Ultimate compressive strength, MPa	13–14	16–17	8–9
Electrical resistance of specimen cube with 50 mm sides, ohms	2	1.1	0.65
Specific electrical body conductivity, siemens-cm	10	17–18	25

It should be pointed out that the potential possibilities of such polymer concretes are far from being exhausted. In the next two or three years, polymer concretes may be produced with much higher electrical conductivity and strength characteristics.

Polymer concretes resistant to radiations

The ability of materials to maintain their properties after irradiation to a certain level (threshold dose) is called radiation resistance. A measure of this resistance is the threshold dose at which some properties of the material undergo significant variation.

The content of water in cement concretes does not exceed 20–25% of the weight of cement or 3–4% of the weight of concrete while the content of polymer binder in polymer concretes can vary from 8–15% of the total weight of the polymer concrete. Calculations have shown that water and polymer binder are roughly equivalent with respect to the effect of penetrating radiations, suggesting the high effectiveness of polymer concretes as protective materials. Thus, investigations revealed that fine-grained polymer concrete with a density of 3230 kg/m^3 based on polyester resin PN-1 and baryte sand fraction 5 mm is 1.5 times superior to cement concrete in protection against γ-radiations. The concentration of hydrogen nuclei in 1 cm^3 of polymer concrete is 1.5 times more than in cement concrete (0.67×10^{22} and 0.43×10^{22} respectively). Tests in the neutron generator NG-15 m of polymer concretes based on polyester resins and cement concrete grade V45 containing granite aggregates (20% of chemically combined water on the weight of cement) revealed that the weakening of the neutron does with different energies was, on average, 40% more in the case of polymer concretes.

At the same time, as a result of the action of ionising radiations at the atomic and molecular level, qualitative changes occur in the microstructure of the polymer binder. These effects can impair or improve the material characteristics. For example, it is known that radiation polymerisation occurs at certain radiation levels of radioactive Co^{60} on monomers of the type methyl methacrylate. During this polymerisation, the level of cross-linkages, and correspondingly the strength characteristics of the polymer produced, is considerably higher than when adopting thermocatalytic polymerisation. The effect of ionising irradiation on thermoplastic polymers at certain doses also leads to improved strength characteristics as a result of additional cross-linkages and the appearance of cross-bonds in the polymer.

The high content of hydrogen even in thermosetting polymers predetermines the degree of their radiation resistance. Significant changes do occur in these materials at high radiation doses.

Investigations on radiation resistance of polymer concretes based on furan and epoxide resins were carried out at the All-Union Order of Lenin S.Ya. Zhuk Design and Research Institute and Order of Red Banner of the Labour V.V. Kuibyshev Moscow Engineering and Design Institute in gamma-ray units, proton accelerators and linear electron accelerator. In the gamma units, the radiation source was radioactive Co^{60} with doses ranging from 1^{-2} to 555^{-2} GR/sec; an irradiation temperature of 20–30°C was maintained by a special cooling system. Radiation loads on the samples during irradiation were determined using dosimetric systems of total radiation absorption.

Radiation resistance was evaluated from the change of elastic and strength characteristics after irradiation at different doses compared to the control specimens. Prisms 40 mm $\times$ 40 mm $\times$ 160 mm were tested for compression and prism strength, modulus of elasticity and Poisson's coefficient.

It was established that when polymer plasters based on FAM resin were irradiated at doses up to 5×10^7 GR, their strength did not vary while the modulus of elasticity rose roughly by 50%. Polymer plasters based on epoxide resin at irradiation doses up to 10^7 GR recorded a 38% strength reduction and at 5×10^7 GR, 70%, but the modulus of elasticity doubled.

The increase in modulus of elasticity and brittle strength of FAM polymer plasters is associated with the additional cross-linkage of the polymer. The much higher increase in modulus of elasticity and reduction of strength of polymer plasters based on epoxide binder are explained by the more intense cross-linkage of the polymer and the simultaneous commencement of its degeneration.

The variation pattern of strength and modulus of elasticity of polymer concretes on irradiation is the same but the strength and deformation properties are stabler since the greater extent of filling with coarse fractions of aggregates promotes radiation resistance.

Investigations established the threshold of radiation damages: for polymer concretes based on ED-20 resin $(1 \text{ to } 2) \times 10^6$ GR and for polymer concretes based on FAM resin $(2 \text{ to } 3.5) \times 10^6$ GR.

The radiation resistance of polymer concretes can be significantly enhanced by using heavy, especially barium-containing aggregates and metallic inclusions in their compositions.

Polymer concretes with high damping and other special properties

It is known that one of the major drawbacks of cast iron is the long duration of the technological cycle and the amount of labour involved in producing cast iron products. The technological process includes preparation of the model, making the earthen mould, casting, breaking the mold, treatment of the casting by shot blasting, rough mechanical processing, thermal treatment, final mechanical processing, priming, puttying, first coat painting, polishing and final painting. The power consumption is very high in the melting of cast iron. After casting, a prolonged process of structural stabilisation occurs in the cast iron which necessitates prolonged thermal treatment or prolonged natural aging. Many companies cure the cast iron frames for up to two years.

The damping capacity of cast iron is low and causes considerable difficulties when producing precision machines. The high thermal conductivity and low thermal capacity of cast iron are responsible for its high sensitivity to the variation of geometric dimensions under the influence of external temperature variations.

Steel began being used as a structural material roughly in the 1950s. Its main advantages are high rigidity of the welded frames made of steel which, compared to frames of cast iron of the same rigidity, are roughly half as heavy, and the high mechanical strength and low coefficient of thermal deformation.

However, steel has some definite disadvantages: low damping characteristics; high friction coefficient as a result of which cast iron or other materials have to be used for making the guides; and the prolonged welding cycle of frames made of steel members. Further, the welded joints cause high internal stresses and sometimes even warping of the welded members and the frame as a whole; low coefficient of resistance to water and cutting fluids; and the long-term dimensional stability of welded structures is inferior to that of cast iron.

Natural stone began being used in the last decade for producing control (gauge) and base plates of machinery in the electronic industry. Such stones, high-strength granites, gabbro-diabases, etc., possess a very high damping capacity (roughly three times more than that of cast iron and steel), high mechanical strength, low deformability and high wear resistance. Natural stone materials have undergone prolonged natural aging and internal stresses are practically absent in them. These together with the low coefficient of temperature deformation ensure high dimensional stability. All of these properties offer much interest for their use in machinery fabrication.

Nevertheless, natural stone material suffers from some characteristic deficiencies: year after year, stone material has become increasingly more expensive and scarce since locating and quarrying large blocks without defects has become problematical; and sawing and mechanical processing leave a large amount of wastage (up to 60%). Instances are not wanting when internal defects turn up in the final stages and almost the whole of the finished product has to be rejected;

working up to the required dimensions, polishing and grinding involve much labour and time; most of the natural rocks adsorb moisture to some extent or the other which adversely affects their dimensional stability.

Cement concrete was first used in the USA in 1866 in metal cutting lathes but its use faded out subsequently. During World War I, it was used for economising metal. In the course of further developments, a small number of machines used various members made of reinforced concrete.

The use of cement concretes as a structural material for body components received a fresh impetus in the early 1860s [*sic*]. Reinforced concrete is used in some quantity for producing large turning, drilling, cutting and other machines in the USSR, the USA, Switzerland, the FRG, France, England and China.

The main advantage of cement concrete is its cheapness, comparatively simple production process and very high damping characteristics compared to cast iron and steel. Examples are known of using prestressed reinforced concrete, fibre concrete and also cement concrete impregnated with polymers.

However, most researchers and specialists agree that cement concrete holds no future for producing base components, precision machines and machines for the radio and electronic industry.

Among the deficiencies of reinforced concrete are low mechanical strength, especially bending and tension, and considerable shrinkage during hardening; comparatively low dimensional stability of cement concrete—even the protection of the concrete surface by many layers of paint does not guarantee dimensional stability; and prolonged hydration of cement (up to 40 days or more).

Polymer concretes are not only free from most of the above drawbacks but also possess several additional positive properties.

The Swiss firm Fritz Studer was one of the first companies in the world to perfect the production of grinding machines with digital programming using polymer concrete frames. In 1974, it commenced extensive research on the use of polymer concretes in machine tool manufacture.

In collaboration with the chemicals producing company Ciba-Geigy, work was carried out in developing polymer concretes 'Granitan s-100' based on epoxide resin Araldite. Granite rubble and sands of various size composition were used as aggregates. Many companies, including those in the FRG, the USA and England, bought the licences and the technology developed by this company.

From the commencement of commercial production (1976) through 1986, over a thousand machines with frames made of polymer concrete were produced in the works of the Studer company in Switzerland and other countries under its licence. The company plans to raise the output of such machines by 50% of the total produced.

Much independent research is being carried out in the use of polymer concretes in machine tool manufacture in the FRG, the USSR, the USA, England, Japan and other countries.

In the FRG major research on polymer concretes based on acrylic and polyester resins is being carried out at the Darmstadt Technical School of the Institute of Prestressed Technology and Machinery, Rheine-Westfalen Higher Technical School (Aachen), and at the major machine tool manufacturing works (Emag, Elb-Shlif, etc.).

By 1983, some 60 machines had been built in the FRG using acrylic polymer concrete members, which rose to 200 by 1984. Emag Company completed the construction of a highly automated plant and organised the commercial production of pointed journals of lathes using polymer concrete members based on epoxide resins.

Various companies specialise in the production of machines using polymer concrete base components: pointed journals of lathes—Emag and Index; grinding machines—Elb-Shlif; milling machines—SKF and Kalle; and co-ordinating and boring mills with digital programming—Burkhat and Weber. More than ten companies are producing gauge plates. The volume of production by all these companies is 3000 to 5000 tonnes a year.

A plant is being built at Crenfield University in England for producing machine components made of polymer concrete with a capacity of 2000–3000 tonnes a year. Orders are on hand for producing machines weighing up to 12 tonnes for turning lathes and journals. In the USA, similar polymer concretes are used by Cincinnati Milacron, Zhendis Flikhslain, etc. Similar work is being carried out in Japan on epoxide and polyester resins at the Tokyo Institute of Technology, universities of Kobe and Toyohashi, and the Technical Research Institute of the Japanese Society of Engineering Co-operatives. The Toyoda Machine Tools has produced a test series of machines (Fig. 11.14).

The main advantage of polymer concretes as material for the base members of machines is their high damping capacity—6 to 10 times more than that of cast iron and steel. As a result, the accuracy and cleanliness of making the components can be improved by 2 to 2.5 times and tool stability on increasing the cutting speed to 40 to 50% by 2 to 2.5 times. Further, among the positive characteristics of polymer concretes are: low thermal conductivity (20 times less than that of cast iron) and high thermal capacity; high stability to the action of water and cutting fluids; high mechanical characteristics under tension making possible the production of frames without reinforcement or using reinforcement only for some heavily stressed components; high dimensional stability; and very simple technology for producing the base components.

It is known that the cost of polymer concretes is more than that of cast iron but experience has shown that polymer concrete frames not only improve the technical characteristics of the machine, but also reduce its cost compared to similar ones made of cast iron or steel, as a result of reduced duration and labour of fabrication. Commercial production of polymer-concrete-based components may reduce the expenses by 50% and electrical power consumption by 40% compared to metal-based machines.

Fig. 11.14. Frame of turning lathe made of polymer concrete based on epoxide resin (FRG).

It should be pointed out that the noise level when operating machines with polymer concrete frames is much less compared to that of similar ones made of cast iron or steel.

Among the deficiencies of stone materials are their great shortage (many western European governments import them from South Africa, India and other countries), high labour and low coefficient of rational utilisation. In many cases, wastes run up to 60% of the weight of the rock.

Compared to natural stones, polymer concretes have the following advantages in this field: less shortage and processing labour, very high damping characteristics and very simple production technology.

Laboratory tests carried out at the Darmstadt Technical School showed that the hardness, wear resistance and chemical stability of polymer concretes based on MMA resin are roughly the same as of granite while the planeness of processed panels 400 mm × 400 mm is about 6 μm. As a result of the high degree of filling with mineral aggregates, the coefficient of temperature deformation (CTD) of polymer concrete is slightly more than that of granite. The cost of polymer concrete gauge plates is 30 to 50% less than for those made from natural stone (Fig. 11.15).

Studies are being carried out on producing supports, tool holders and other components of metal processing machines using such concretes.

In the Soviet Union, work on the use of polymer concretes in lathe and

294

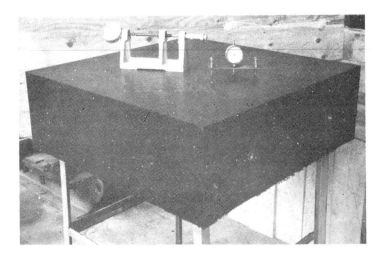

Fig. 11.15. Gauge plate 800 mm × 800 mm × 250 mm made of polymer concrete based on FAED compound.

machine fabrication is being carried out at the Reinforced Concrete Research Institute and Experimental-Research Engineering Bureau in collaboration with several design and production organisations. Table 11.9 gives the average values of the properties of materials used in lathe and machine fabrication.

Since one of the advantages of polymer concretes is their high damping property, i.e., the ability to disperse and absorb mechanical oscillations, great attention is being paid to this property.

It is known that the damping properties of materials is an independent characteristic which depends little on the other mechanical properties of a given material. Among the basic factors influencing the damping of oscillations are: amplitude and frequency of cyclic oscillations; type of oscillations (longitudinal, transverse, torsional, or their combinations); dimensions and shape of the article; structure and temperature of the material (raising the temperature increases the damping property); duration of the action of cyclic stresses and creep of the material (the higher the creep, the greater the absorption of oscillation energy).

According to the presently popular theory of plastic deformations, oscillation damping occurs as a result of plastic deformations in the material. However, solutions to the applied problems based on empirical equations provide only very approximate values.

According to the modern concepts, microscopic deformations represent the basic, but not the only, source of scattering of the energy. A significant part of the vibration process is expended in heating the material. In combination systems such as polymer concrete, the scattering of energy is also caused by the movement of micro- and macromolecules of the polymer which generates

Table 11.9. Properties of materials used in lathe and machine fabrication

Material	Density, kg/m^3	R_{comp} MPa	R_b, MPa	E, MPa $\times$ 10^4	CTD, °C $\times$ 10^{-6}	Thermal conductivity, W/(m·K)	Thermal capacity, kJ/(kg·K)	Coefficient of oscillations damping
Gray cast iron	7200–7500	400–900	160–400	9.5–11.7	9–12	60–75	0.5–0.8	1
Mild steel	7800	400–500	460	20	9–12	45–60	0.46	0.8
Granite	2600–2800	100–240	10–20	3.8–76	4.7–82	3–4	0.85–0.95	3
Cement concrete	2400–2500	40–50	6–8	3.0–3.5	10–12	0.8–1.3	0.8–0.9	4–5
Polymer concrete 'Granitan' (Switzerland)	2200–2400	100–120	20–35	3.6–4.2	12–14	0.8–1.9	1.25	6–6.5
Polymer concrete based on acrylic resin 'Motema' (FRG)	2200–2400	115–150	29–34	3.0–3.8	17	2	—	6–7
FAED polymer concrete based on furan-epoxide resin (USSR)	2200–2400	120–140	30–35	3.7–3.9	10–14	0.8–0.9	—	6–8

additional zones of friction. Filled polymers are also characterised by elastic-viscous scattering of energy similar to the losses of energy during the oscillations of solid bodies in viscous liquids, which are described by Foigt's theory.

Adequate data are available in the literature on the damping properties of various materials, primarily metals. At the same time, similar data for polymer concretes are extremely scant.

The logarithmic decrement of oscillations damping (δ) is taken as a quantitative characteristic of the damping capacity of a material. It is equal to the ratio of the energy scattered in one cycle to double the energy of the cycle:

$$\delta = \Delta W/2W. \qquad \ldots (11.2)$$

This equation is relevant only in the absence of a relation between frequency and scattering of energy, e.g., for metals. For polymer materials, this dependence is relevant in the resonance zone.

In practice, δ is derived by processing the vibrograph of free oscillations:

$$\delta = (1/n)\ln[a_i/(a_{i+n})], \qquad \ldots (11.3)$$

where a_i and a_{i+n} are, respectively, the amplitudes at the commencement and end of the period comprising n cycles.

During cyclic oscillations, δ is found in the resonance zone:

$$\delta = (\pi/\sqrt{3})(\Delta f_{0.5}/f_0) = \pi(\Delta f_{0.7}/f_0), \qquad \ldots (11.4)$$

where f_0 is the natural frequency, Hz; $\Delta f_{0.5}$ the width of resonance cycle at half its height, Hz; and $\Delta f_{0.7}$ the same at half the level.

The decrement of oscillations damping was determined at the Reinforced Concrete Research Institute using frequency-damping meter type Ig 3-3410. The device helps to determine the resonance frequency of oscillations in a specimen from the maximum amplitude with digital display of data in the frequency range 100 Hz to 25 kHz. It automatically determines the value of δ if it falls in the range 0.006 to 0.3.

Comparative tests on various materials were carried out on specimens 40 mm × 62 mm × 400 mm (Table 11.10). The damping decrement of polymer concretes based on MMA and FAED resins is 2 to 2.5 times more than that of polymer concretes based on epoxide resins, 5 times that of cement concrete, 11–13 times that of gabbro-diabase, and 70–80 times that of steel.

Very high decrement values greatly restrict the oscillations amplitude. In the range of frequencies that differ from resonance frequencies, the amplitude restriction is not so significant but, on average, the amplitude for polymer concretes is 6–10 times less than that of steel.

It should be pointed out that there is an optimum zone of binder content for polymer concretes which makes for maximum decrement for a given composition.

Table 11.10. Results of comparative tests on various materials

Material	Natural frequency f_0, Hz	Logarithmic decrement δ	Resonance amplitude (amplitude for steel adopted as 1000 units)
Hardened epoxide resin ED–20	441.3	0.064	13.1
Hardened furan-epoxide resin FAED	420.13	0.075	11.1
Polymer concrete based on MMA binder	890.1	0.063	13.3
Polymer concrete based on furan-epoxide binder FAED	973.4	0.067	12.5
Composite beam—polymer concrete FAED with steel bar 7 mm thick moulded on it	1470	0.063	13.3
Polymer concrete based on epoxide binder of firm Studer (company data)	845	0.0206	–
Polymer concrete based on epoxide binder ED–20	903	0.027	31
Wood (pine)	1290	0.023	36.3
Cement concrete grade V30, W/C 0.44	955.7	0.012	69
Gabbro-diabase	1589	0.005	167.2
Steel (St3)	1277.6	0.0008	1000

The zones of maximum damping decrements almost coincide with the zones of optimum compositions of polymer concretes in respect of their strength characteristics.

Polymer concretes with high thermal stability

It is known that one of the main drawbacks of polymer concretes is their low thermal stability (mainly 80–150°C). This value restricts in many cases the field of application of these progressive materials. Therefore, research is being carried out in many countries for developing new types of polymer concretes with very high thermal stability. For example, the Brookhaven National Laboratory (USA) has developed new types of thermally resistant polymer concretes which underwent laboratory and field tests in highly concentrated salt solutions and hot salt solutions of the same concentrations up to 300°C.

Table 11.11 presents the results of strength variation of polymer concretes depending on the composition and retention periods in the aggressive medium at 237°C. All the five types of polymer concretes contained 13% complex binder and 87% aggregates by weight. Tests showed that the strength of all the compositions rose in the course of 30 days compared to the control specimens, as a result of additional polymerisation under the action of the high temperature of the salt solution. On further retention, the strength decreased steadily but, after 240 days of tests, the strength of the fifth composition exceeded that of control,

Table 11.11. Variation of strength of polymer concretes in relation to composition and retention in an aggressive medium

Composition of binder, % by weight				R_{comp}, MPa, after retention in aggressive medium			
Styrene	Acrylo-nitrile	Acryl-amide	Divinyl-benzene	Control specimens	Test period, days		
					30	120	240
52.5	40	5	2.5	191.6	211.9	160.9	124.7
62.5	30	5	2.5	202.4	227	174.9	174.2
50	40	5	5	189.5	196.1	147.3	123.8
50	37.5	5	7.5	176.6	237.1	177.8	162.6
50	35	5	10	180.8	237.6	231.9	204.1

i.e., 204 and 180.8 MPa respectively.

Long-term results of laboratory and field tests under various geothermal conditions confirmed the high efficiency of the developed polymer concrete compositions.

The polymer concretes laboratory of the Reinforced Concrete Research Institute developed a high-temperature composition of polymer concrete based on epoxide-silicone binder and zirconium aggregate for which the short-term thermal stability for several hours was 600°C and long-term 400°C.

Among the special types are the air-tight polymer concretes which at a minimum thickness of 50 mm are capable of maintaining a high vacuum for a long period.

This property is achieved not only by the composition of the polymer concrete but also by the technology of its production since the smaller the number of air inclusions in the concrete structure, the greater its air-tightness, while the amount of air pores is primarily determined by the technological process adopted. Therefore, when producing such concretes, synthetic resins and aggregates heated roughly to 60°C are recommended for use. The mix should be mixed and moulded under vacuum. At size 5 m × 7 m × 12 m and using assembled members, the joints measured about 300 m.

A single-piece moulding of such large units while ensuring uniform strength at a vacuum load of 100 kN/m² and air-tightness of the entire structure was by itself a complex engineering task. A vacuum chamber that underwent production tests has been commissioned.

Concluding remarks

The above list of rational applications of polymer concretes in diverse fields is far from complete. This list naturally could not cover all the various uses of these advanced structural materials which have been steadily growing year after year. At the same time, the attempted systematisation of the application of polymer concretes in diverse fields enables a more tangible evaluation of their importance in construction practice and in the industry and indicates more realistically their further development and application.

It should be pointed out, however, that a wider application of polymer concretes is impeded by the comparatively high costs of monomers, oligomers, and hardening and modifying additives which are naturally reflected in the cost of polymer concrete products and structures. With the improved production technology of synthetic resins and the essential components, and their greater production, some reduction in costs can be anticipated, but the main factor that can reduce the price of polymer concrete is the reduction of the consumption of polymer binder per unit weight of polymer concrete produced. Much has been done along these lines but all the possibilities have not been exhausted and research on reducing the polymer binder consumption should be actively pursued.

Another extremely important factor restraining the application of polymer concretes to some extent, is the non-availability of local and international specifications in many countries.

In the Soviet Union, based on fundamental investigations, the theory of structure formation of polymer concretes and methods of selecting optimum compositions and the theory of designing polymer concrete structures have been developed. Based on theoretical considerations and experience in using different types of structures, more than twenty departmental and governmental instructional-standard documents have been formulated, including Instructions on the Production Technology of Polymer Concretes and Their Products, SN 525–80; Instructions for Designing Structures of Buildings and Installations using Reinforced Polymer Concrete, VSN 12–84; Guide for Designing Polymer Concrete and Reinforced Polymer Concrete Structures Using Stressed and Unstressed Reinforcement, GOST 25881–83, GOST 25246–82, etc., and also albums of type drawings of the more frequently used polymer concrete products and structures.

The USA, France and Japan have several departmental standards on methods of testing polymer concretes. Similar work is being carried out in the FRG where, in addition, work has been carried out on designing structural members, including the bodies of large reducers, but governmental standards relating to polymer concretes have yet to be brought out.

Table 11.12 (compiled by the State Institute for Designing Non-ferrous Metallurgical Plants) presents comparative structural diagrams and technoeconomic indices of reinforced concrete structures with anticorrosive protection and chemically stable structures of polymer concretes. The Table strikingly demonstrates the vital differences in the material content, difficulty of production and the actual economic efficiency of polymer concrete structures compared to conventional structures.

Table 11.12. Comparison of structural features and techno-economic indexes of chemically anticorrosive protection

Structure	Reinforced concrete structure with anticorrosive protection			
	Drawing	Material	Labour, man-h	Cost, '000 roubles
Foundation of building frame		1–precast reinforced concrete foundation (8.2 m^2); 2–waterproofing with asphalt concrete δ = 40 mm (12.6 m^3); 3–waterproof backing in two coasts of polyisobutylene (39.5 m^2); 4–protective layer of cement mortar δ = 20 mm (10.8 m^2); 5–acid-resistant aggregate filled with bitumen to saturation δ = 100 mm (12.6 m^2); 6–impervious soil (12.6 m^2); 7–clay bricks on bitumen (30.8 m^2)	269	1.59
Foundation beams		1–reinforced concrete foundation of beam (0.43 m^3); 2–3 coats of hot bitumen on cold soil (5.2 m^2)	7.3	0.05

stable steel polymer concrete structures with reinforced concrete structures given

Chemically stable structure		Labour, man-h	Cost, '000 roubles	Annual economy, '000 roubles	
Drawing	Material			Structural member	Copper electrolysis shop
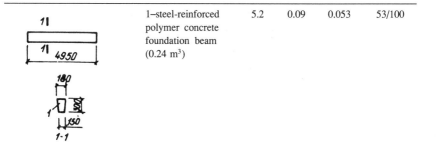	1–steel-reinforced polymer concrete foundation (5.6 m³); 2–underlayer of sand $\delta = 100$ mm (7.2 m²); 3–impervious soil (0.7 m³)	13	1.95	1.73	429/248
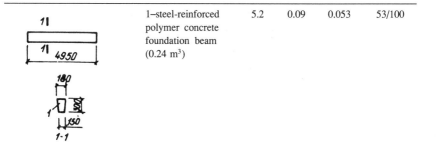	1–steel-reinforced polymer concrete foundation beam (0.24 m³)	5.2	0.09	0.053	53/100

302

(Table 11.12. *contd.*)

Structure	Reinforced concrete structure with anticorrosive protection			
	Drawing	Material	Labour, man-h	Cost, '000 roubles
Wall panels of basement	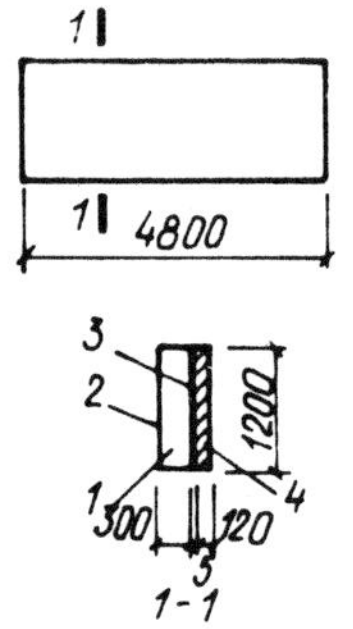	1–reinforced concrete panel of basement (1.73 m³); 2–lining with ceramic tiles using acid-resistant mortar (5.8 m³); 3–waterproof backing in two coats of poly-isobutylene (5.8 m²); 4–clay bricks on bitu-men (5.8 m²)	57.6	0.43
Columns of building frame	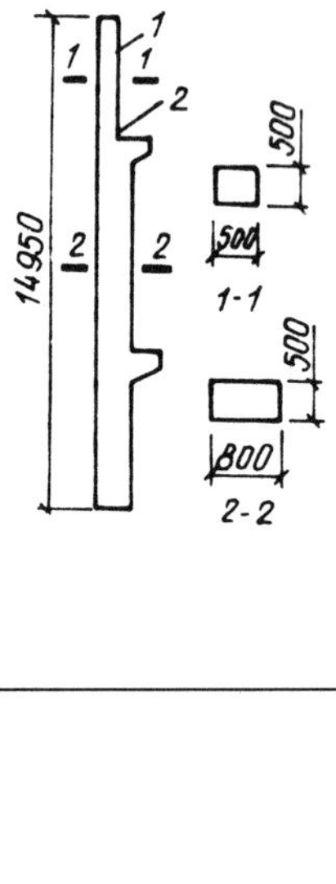	1–reinforced concrete columnar frame of building (5.7 m³); 2–4 coats of chemi-cally resistant enamel on two coats of che-mically resistant lac-quer (36 m³)	61.4	0.87

Chemically stable structure				Annual economy, '000 roubles	
Drawing	Material	Labour, man-h	Cost, '000 roubles	Structural member	Copper electrolysis shop
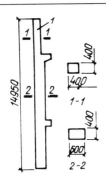	1–steel-reinforced polymer concrete wall panel of basement (1.1 m³)	5.6	0.42	0.53	158/300
	1–steel-reinforced polymer concrete column of building frame (3.4 m³)	13.6	1.51	0.84	209/248

(Table 11.12. *contd.*)

Structure	Reinforced concrete structure with anticorrosive protection			
	Drawing	Material	Labour, man-h	Cost, '000 roubles
Bath supports		*Structure* 1–reinforced concrete beam (8 + 4 numbers; 15.6 m^3); 2–reinforced concrete column (16 numbers; 13 m^3); 3–reinforced concrete foundation (16 numbers; 24.5 m^3)	269	7.2
		Anti-corrosive treatment 4–2 coats of fibreglass in epoxy resin (402 m^3); 5–acid-resistant bricks in $\frac{1}{2}$ bricks on andesite mortar (17 m^2); 6–holding wall of red brick on bitumen (124 m^2); 7–priming with bituminol (124 m^2).	122	5.52
		Total for the above.	149	12.8
Baths		1–lead lining (1.21 tonnes); 2–3 coats of bitumen lacquer (18 m^2);	43.4	1.17
		3–4 coats of EP–0010 primer (21.6 m^2); 4–reinforced concrete bath wall (1 number); 5–2 coats of fibreglass in epoxy resin ED–5 (21.6 m^2)	69.6	0.6
		Total for 1 number.	113	1.77

Chemically stable structure				Annual economy, '000 roubles	
Drawing	Material	Labour, man-h	Cost, '000 roubles	Structural member	Copper electrolysis shop
	1–steel-reinforced polymer concrete beam (8 numbers; 13.7 m^3); 2–steel-polymer concrete internal member (3 numbers; 8.6 m^3); 3–steel-reinforced polymer concrete support (6 numbers; 4.6 m^3). Total for the above.	109	9.92	16	640/40
	Variant 1 1–steel-reinforced polymer concrete bath (1 number)	23.2	0.88	2.32	2420/1040
	Variant II 2–steel-reinforcement polymer-silicate bath (1 number)	23.2	0.56	2.42	2510/1040

(Table 11.12. *contd.*)

Structure	Reinforced concrete structure with anticorrosive protection			
	Drawing	Material	Labour, man-h	Cost, '000 roubles
Built-in stack		1–reinforced concrete column (1.41 m³); 2–reinforced concrete plates (2 numbers); 3–4 coats of chemically resistant enamel on 2 layers of chemically resistant lacquer (25.2 m²); 4–levelling plate of cement mortar $\delta = 20$ mm (18 m²); 5–2 coats of waterproof backing of polyisobutylene (10 m²); 6–acid-resistant bricks in $\frac{1}{4}$ bricks on andesite mortar (18 m²)	137	0.72
Chemically resistant floors		1–acid-resistant bricks in $\frac{1}{4}$ bricks on bituminol paste (1 m²); 2–acid-resistant bricks in $\frac{1}{2}$ bricks on andesite mortar (1 m²); 3–priming with andesite mortar $\delta = 5$ mm (1 m²); 4–PSG polyisobutylene $\delta = 2.5$ mm in 2 coats on 88–N clay (1 m²); 5–levelling layer of cement-sand mortar $\delta = 20$ mm (1 m²); 6–floor base (1 m²)	11.8	0.04

	Chemically stable structure			Annual economy, '000 roubles	
Drawing	Material	Labour, man-h	Cost, '000 roubles	Structural member	Copper electrolysis shop
	1–steel-polymer concrete column (1.23 m³); 2–steel polymer concrete plate covering (2.34 m²)	15.9	1.22	0.71	628/800
	Variant I 1 — polymer concrete plate (1 m²); 2 — polymer paste δ = 10 mm (1 m²); 3 — polymer-silicate concrete δ = 80 mm (1 m²); 4 — PSG polyisobutylene δ = 2.5 mm in 2 coats on 88–N clay (1 m²); 5–levelling layer of cement-sand mortar δ = 20 mm (1 m²); 6–floor base (1 m²)	8.98	36.9	0.06	1121/20,000
	Variant II 7–plasticised polymer paste δ = 3 mm (1 m²); 8–polymer-silicate concrete δ = 80 mm (1 m²)	7.9	31.2	0.06	1140/20,000

(Table 11.12. *contd.*)

Structure	Reinforced concrete structure with anticorrosive protection			
	Drawing	Material	Labour, man-h	Cost, '000 roubles
Wall panels		1–wall panel in 2 layers (1st layer —heavy concrete ρ = 2500 kg/m³; 2nd layer–light concrete ρ = 1200 kg/m³; 1 number); 2–4 coats of polyvinylchloride on 2 coats of chemically resistant lacquer primer (10.8 m²); 3–4 coats of chemically resistant enamel on 2 coats of chemically resistant lacquer primer (10.8 m²)	9.32	0.29
Foundation for equipment		1–reinforced concrete foundation for equipment (0.97 m³); 2–acid-resistant bricks in $\frac{1}{4}$ bricks on andesite mortar (5.5 m²); 3–2 coats of waterproof backing with polyisobutylene (4.9 m³); 4–acid-resistant plaster filling under the frame of equipment δ = 30 mm (1.44 m²)	43.5	0.22

* Total annual economy was calculated for polymer concrete structures (variant I).

Chemically stable structure				Annual economy, '000 roubles	
Drawing	Material	Labour, man-h	Cost, '000 roubles	Structural member	Copper electrolysis shop
 1-1	1–wall panel of perlite polymer concrete ρ = 500 kg/m^3 (1.08 m^3)	10.3	0.27	0.35	202/574
	Variant I 1–steel-reinforced polymer concrete foundation (0.4 m^3); 2–fixing of bolts in epoxide adhesive (6 kg); 3–polymer plaster filling under frame of equipment δ = 30 mm (1.1 m^2)	2.72	0.12	0.29	86.94/300
	Variant II 1–steel-reinforced polymer silicate foundation (0.4 m^3); 2–fixing bolts in epoxide adhesive (6 kg); 3–polymer-silicate plaster filling under frame of equipment δ = 30 mm (1.08 m^2)	2.72	0.08	0.3	90.36/300
	Total		6883.4*		

Literature Cited*

1. Agadzhanov, V.I. 1976. Ekonomika povysheniya dolgovechnosti i korrozionnoi stoikosti stroitel'nykh konstruktsii [Economics of Improving the Durability and Corrosion Resistance of Building Structures]. Stroiizdat, Moscow, 112 pp.

2. Akhverdov, I.N. *et al.* 1973. Modelirovanie napryazhennogo sostoyaniya betona i zhelezobetona [Modelling the Stress State of Concrete and Reinforced Concrete]. Minsk, 231 pp.

3. Akutin, M.S. *et al.* 1974. Novye polimernye materialy s ponizhennoi vozgoraemost'yu na osnove polistirola i polietilena [New polymer materials with reduced combustibility based on polystyrene and polyethylene]. *Plasticheskie Massy*, No. 12, pp. 36–37.

4. Al'shits, I.M. 1964. Poliefirnye stekloplastiki dlya sudostroeniya [Polyester Glass Plastics in Ship Building]. Sudostroenie, Leningrad, 288 pp.

5. Andrievskaya, G.D. 1967. Fiziko-khimiya i mekhanika orientirovannykh stekloplastikov [Physical Chemistry and Mechanics of Oriented Glass Plastics]. Nauka, Moscow, pp. 3–14.

6. Askadskii, A.A. and G.L. Slonimskii. 1965. *Mekhanika Polimerov*, No. 4.

7. Balalaev, G.L., V.M. Medvedev and N.A. Moshchanskii. 1966. Zashchita stroitel'nykh konstruktsii ot korrozii [Corrosion Protection of Building Structures]. Stroiizdat, Moscow, 224 pp.

8. Bares, R. 1982. Klassifizierung von Komposit-Werkstoffen und die Kompositen mit Plasten. *Kunststoff im Bau*, **17**, No. 1.

9. Bares, R. Relation between geometric and physical structure and properties of granular composites. *J. Material Science* (in press).

10. Barten'ev, G.M. and Yu.K. Zuev. 1974. Prochnost' i razrushenie vysokoelasticheskikh materialov [Strength and Degradation of Highly Elastic Materials]. Khimiya, Moscow, 387 pp.

11. Basin, V.E. and A.A. Berlin. 1970. Problemy adgezionnoi prochnosti [Problems of adhesive strength]. *Mekhanika Polimerov*, No. 2, pp. 303–310.

12. Bazhenov, V.A. *et al.* 1972. Problemy biologicheskikh povrezhdenii i obrastaniya materialov, izdelii i sooruzhenii [Problems of Biological

---

* Some entries were incomplete in Russian original—General Editor.

Degradations and Overgrowths on Materials, Products and Equipment].
Nauka, Moscow.

13. Berg, O. Ya. 1950. K voprosu o prochnosti i plastichnosti betona [Aspects of strength and plasticity of concrete]. *Dokl. AN*, **70**, No. 4, pp. 617–620.

14. Berlin, A.A. *et al.* 1967. Poliefirakrilaty [Polyester Acrylates]. Nauka, Moscow, 472 pp.

15. Borodkina, N.I., E.A. Vakhtangova *et al.* 1971. Poluchenie, primenenie i modifitsirovanie ATsF smol [Production, application and modification of ATsF resins]. *Plasticheskie Massy*, No. 4.

16. Busev, A.I. and L.N. Simonova. 1975. Analiticheskaya khimiya sery [Analytical Chemistry of Sulphur]. Nauka, Moscow, 272 pp.

17. Chebanenko, A.I. 1970. Issledovanie napryazhenno deformirovannogo sostoyaniya nesushchikh i polimerbetonnykh konstruktsii pri pomoshchi ob''emlyushchikh diagramm [A study of the stress-deformed state of load-bearing and polymer concrete structures using enveloping diagrams]. In: *Konstruktivnye i khimicheski stoikie polymerbetony.* Stroiizdat, Moscow, pp. 54–69.

18. Chebanenko, A.I. 1972. Osnovy rascheta stalepolimerbetonnykh konstruktsii [Principles of designing steel-polymer concrete structures]. In: *Stalepolimerbetonnye stroitel'nye konstruktsii.* Stroiizdat, Moscow, pp. 121–150.

19. Chebanenko, A.I. 1980. Reologicheskie svoistva armirovannogo polimerbetona [Rheological properties of reinforced polymer concrete]. In: *Issledovanie stroitel'nykh konstruktsii s primeneniem polimernykh materialov.* VPN, Voronezh, pp. 7–14.

20. Chebanenko, A.I. 1984. Osnovy teorii rascheta armopolimerbetonnykh konstruktsii [Principles of the theory of designing reinforced polymer concrete structures]. *Beton i Zhelezobeton*, No. 8, pp. 5–8.

21. Chebanenko, A.I. *et al.* 1980. Reologicheskie svoistva armirovannogo polimerbetona [Rheological properties of reinforced polymer concrete]. In: *Issledovanie stroitel'nykh konstruktsii s primeneniem polimernykh materialov.* VPN, Voronezh, pp. 7–14.

22. Chekhov, A.P. 1977. Zashchita stroitel'nykh konstruktsii ot korrozii [Protection of Steel Structures against Corrosion]. Vysshaya Shkola, Kiev, 214 pp.

23. Choshchshiev, K.Ch. 1983. Tekhnologiya polimerbetonov s ispol'zovaniem barkhannykh peskov [Technology of Polymer Concretes using Barkhan Sands]. Ylym, 329 pp.

24. Chuiko, A.V. 1978. Organogennaya korroziya [Organogenic Corrosion]. Saratov State University, Saratov, 230 pp.

25. Czarnecki, L. 1983. Polimer-Beton-Verbundbaustoffe, *Kunststoff im Bau*, No. 4, pp. 33–37.

26. Czarnecki, L. 1984. Untersuchung über den Aufbau von Polimerbeton (Micro und Macrostruktur). In: *Virter Internationler Kongress "Polimer und Beton"*. BRD, Darmstadt, pp. 59–64.

27. Davydov, S.S., N.A. Moshchanskii, V.V. Paturoev and A.I. Chebanenko. 1972. Khimicheski stoikie konstruktsii iz polimerbetonov [Chemically stable polymer concrete structures]. RILEM "Materialy i konstruktsii", Paris, No. 26, pp. 99–104.

28. Desov, A.E. 1966. Nekotorye voprosy struktury, prochnosti i deformativnosti betonov [Some aspects of structure, strength and deformability of concretes]. In: *Struktura, prochnost' i deformativnost' betonov*. Moscow, pp. 4–58.

29. Elshin, I.M. 1980. Polimerbetony v gidrotekhnicheskom stroitel'stve [Polymer Concretes in Hydraulic Engineering Constructions]. Stroiizdat, Moscow, 192 pp.

30. Fantalov, A.M. and V.V. Paturoev. 1979. Vysokomekhanizirovannoe izgotovlenie polimerbetonnykh konstruktsii [Highly mechanised fabrication of polymer concrete structures]. *Beton i Zhelezobeton*, No. 8, pp. 16–18.

31. Fedorov, V.S. 1980. Armopolimerbetonnye konstruktsii povyshennoi ognestoikosti [Reinforced polymer concrete structures with high heat stability]. Avtoref. dissert. kand. tekhn. nauk., Moscow, 22 pp.

32. Fridman, V.V. 1970. Statisticheskii metod v issledovanii sostava poliefirnogo polimerbetona [Statistical method for studying polymer concrete mixes based on polyesters]. *Stroitel'nye Materialy*, No. 6 pp. 28–29.

33. Grassi, N. 1967. Termicheskaya destruktsiya [Thermal degradation]. In: *Khimicheskie reaktsii polimerov*. Edited by E. Fettes, Mir, Moscow, vol. 2 (translated from English).

34. Griffith, A. 1921. *Phil. Trans. Soc.*, **221**.

35. Rukovodstvo po proektirovaniyu polimerbetonnykh i armopolimerbetonnykh konstruktsii s napryagaemoi i nenapryagaemoi armaturoi [Guide for Designing Polymer Concrete and Reinforced Polymer Concrete Structures with Stressed and Unstressed Reinforcement]. TsNIItsvetmet Ekonomiki i Informatsii, Moscow, 1986, 185 pp.

36. Gul', V.E. and V.N. Kuleznev. 1972. Struktura i mekhanicheskie svoistva polimerov [Structure and Mechanical Properties of Polymers]. Vysshaya Shkola, Moscow, 320 pp.

37. Il'yushin, A.A. and A.M. Ogibalov. 1966. O kriterii dlitel'noi prochnosti polimerov [Criteria for long-term strength of polymers]. *Mekhanika Polimerov*, No. 6, pp. 828–832.

38. Instruktsiya po proektirovaniyu i izgotovleniyu bakovoi apparatury iz armopolimerbetona [Instructions for Designing and Fabrication of

Reinforced Polymer Concrete Bath Equipment]. VSNO1–78/MTsM SSSR. Tsvetmetinformatsiya, 1979, 94 pp.

39. Instruktsiya po proektirovaniyu konstruktsii zdanii i sooruzhenii iz armopolimerbetona [Instructions for Designing Reinforced Polymer Concrete Structures of Buildings and Installations].

40. Instruktsiya po tekhnologii prigotovleniya polimerbetonov i izdelii iz nikh [Instructions on Production Technology of Polymer Concretes and Products Therefrom]. SN 525–80. Stroiizdat, Moscow, 1981, 23 pp.

41. Japanese (English)-Russian Glossary of Polymers in Concrete. College of Engineering, Nihon University. Koriyama, Fucku-Shimaken, Japan, 1985, 18 pp.

42. Kapyscinski, J. and K. Pucilowski. 1981. Designing and Technology of Composite Materials. Warsaw Technical University, Warsaw, 32 pp.

43. Kargin, V.A. and G.L. Slonimskii. 1967. Kratkie ocherki po fiziko-khimii polimerov [Brief Outline of the Physical Chemistry of Polymers]. Khimiya, Moscow, 227 pp.

44. Kiselev, M.S., L.A. Sukhareva, V.V. Paturoev and P.I. Zubov. 1967. Fiziko-khimiya i mekhanika orientirovannykh stekloplastikov [Physical Chemistry and Mechanics of Oriented Glass Plastics]. Nauka, Moscow, pp. 187–194.

45. Krylov, L.M. et al. 1967. Issledovanie vliyaniya vzaimodeistviya na granitse polimer-tverdoe telo na mekhanicheskie svoistva alkidnykh pokritii [Study of the effect of reactions at the polymer-solid body boundary on the mechanical properties of alkyd coatings]. Mekhanika Polimerov, No. 1, pp. 19–23.

46. Lexique Terminologique des Betons Avec Resines Synthetiques Francais-Russe Imprime an Laboratoire Central des Ponts et Chaussees. Paris, 1978, 30 pp.

47. Lipatov, Yu. S. 1977. Fiziko-khimiya napolnennykh polimerov [Physical Chemistry of Filled Polymers]. Khimiya, Moscow, 304 pp.

48. Manson, I.A. 1981. Overview of current research on polymer concrete. ASI-Sp-69, pp. 1–17.

49. Menkovskii, M.A. and V.T. Yavorskii. 1985. Tekhnologiya sery [Technology of Sulphur]. Khimiya, Moscow, 327 pp.

50. Moshchanskii, N.A. and V.V. Paturoev. 1970. Konstruktivnye i khimicheski stoikie polimerbetony [Structural and Chemically Stable Polymer Concretes]. Edited by V.V. Paturoev. Stroiizdat, Moscow, 205 pp.

51. Moshchanskii, N.A. and V.V. Paturoev, eds. 1971. Structural and Chemically Stable Polymer Concretes. Jerusalem, 143 pp.

52. Mustavaev, R.M., L.G. Kulieva et al. 1986. Nepredel'nye kremniior-ganicheskie soedineniya kak modifikatory epoksidnoi smoly ED-20 [Un-

saturated organosilicon compounds as modifiers of epoxide resin ED-20].
Plasticheskie Massy, No. 12, p. 49.

53. Obolduev, A.T. 1980. K voprosu povysheniya termoustoichivosti polimerbetonnykh konstruktsii [Aspects of improving the thermal stability of polymer concrete structures]. *Promyshlennoe Stroitel'stvo*, No. 6, 13–14.

54. Ogibalov, A.M. and Yu.V. Sivorova. 1965. Mekhanika armirovannykh plastikov [Mechanics of Reinforced Plastics]. Moscow State University, Moscow, 479 pp.

55. Okhama, E. [Ohama, Y.]. 1980. Sostoyanie i perspektivy razvitiya polimerbetonov i betonopolimerov v Yaponii [Status and prospects of development of polymer concretes and concrete polymers in Japan]. *Beton i Zhelezobeton*, No. 3, pp. 34–36.

56. Ohama, Y. and K. Demura. 1979. Effect of coarse aggregate on compressive strength of polyester resin concrete. *The Int. J. of Cement Composites*, **1**, No. 3, pp. 111–115.

57. Panasyuk, V.V. 1968. Predel'noe ravnovesie khrupkikh tel s treshchinami [Maximum Equilibrium of Brittle Bodies with Fractures]. Naukova Dumka, Kiev, 246 pp.

58. Patent No. 25074, USA, 1859.

59. Patent No. 4343924, USA, August 10, 1982.

60. Patent No. 4395520, USA, July 26, 1983.

61. Paturoev, M.V., V.G. Sviridov et al. 1983. Termoobrabotka polimerbetonnykh izdelii i konstruktsii [Thermal treatment for polymer concrete products and structures]. In: *Khimicheski stoikie P-betony*. NIIZhB. Stroiizdat, Moscow, pp. 50–53.

62. Paturoev, V.V. 1977. Tekhnologiya polimerbetonov (fiziko-khimicheskie osnovy [Technology of Polymer Concretes (Physical and Chemical Principles)]. Stroiizdat, Moscow, 240 pp.

63. Paturoev, V.V. 1987 Polimerbetony [Polymer Concretes]. Stroiizdat, Moscow, 286 pp.

64. Paturoev, V.V., A.N. Volgushev and Yu.I. Orlovskii. 1985. Sernye betony i betony propitannye seroi [Sulphur Concretes and Concretes Impregnated with Sulphur]. Obz. inform. Ser. 7, No. 1, VNIIIS Gosstroya SSSR, Moscow, 59 pp.

65. Paturoev, V.V. and S.Z. Sarnitskaya. 1978. Tsvetnoi polimerbeton dlya pokrytii polov [Coloured polymer concrete for floors]. *Stroitel'stvo i Arkhitektura Uzbekistana*, No. 10, pp. 40–41.

66. Paturoev, V.V. *et al.* 1976. Vozgoraemost' polimerbetonov i ognestoikost' konstruktsii iz nikh [Combustibility of polymer concretes and heat resistance of structures made of them]. *Beton i Zhelezobeton*, No. 3, pp. 25–26.

67. Fiziko-khimicheskie svoistva sery [Physical and Chemical Properties of Sulphur]. Obz. NIITEKhIM, Moscow, 1985, 40 pp.

68. Plastics in Hydrotechnical Construction (Glossary). US Department of Interior, Bureau of Reclamation, Denver, Colorado, 1984, 143 pp.

69. Ramachandran, V., R. Fel'dman and D. Boduen. 1986. Nauka o betone [Science of Concrete]. Edited by V.B. Ratinov. Stroiizdat, Moscow, 278 pp. (translated from English).

70. Shemerdyak, B.M. *et al.* 1977. Antikorrozionnye pokrytiya na predpriyatiyakh kaliinoi promyshlennosti [Anticorrosive Coatings in the Potassium Industry]. Obz. inform. NIITEKhIM, Moscow, 24 pp.

71. Shevchenko, V.I. 1986. Ob otsenke treshchinostoikosti betona po parametram polnykh diagramm izgiba [Evaluating the fracture-resistance of concrete from complete flexure diagrams]. *Zavodskaya Laboratoriya*, No. 3, pp. 64–66.

72. Shlyanskii, O.F. *et al.* 1965. *Plasticheskie Massy*, No. 11.

73. Simonòv-Emel'yanov, I.D. *et al.* 1976. *Plasticheskie Massy*, No. 11.

74. Skupin, L. 1967. Polimernye rastvory i plastbetony [Polymer Plasters and Plastic Concretes]. Stroiizdat, Moscow, 175 pp.

75. Solomatov, V.I., V.I. Klyukin *et al.* 1979. Armopolimerbeton v transportnom stroitel'stve [Reinforced Polymer Concrete in the Transport Industry]. Transport, Moscow, 232 pp.

76. Stalepolimerbetonnye stroitel'nye konstruktsii [Steel-Polymer Concrete Building Structures]. Edited by S.S. Davydov and A.I. Ivanov. Stroiizdat, Moscow, 1972, 280 pp.

77. Struktura i elektrofizicheskie svoistva sery [Structure and Electro-physical Properties of Sulphur]. Obz. NIITEKhIM, Moscow, 1983, 32 pp.

78. Sukhareva, L.A. 1966. Mekhanizm protsessov plenkoobrazovaniya iz polimernykh rastvorov i dispersii [Film Formation Mechanism in Polymer Solutions and Dispersions], Nauka, Moscow.

79. Ur'ev, N.B. and N.V. Mikhailov. 1968. Skleivanie tverdykh poverkhnostei vysokonapolnennymi polimernymi kleyami pri vibratsii [Bonding of hard surfaces using highly filled polymer adhesives under vibrations]. *Stroitel'nye Materialy*, No. 75, pp. 10–11.

80. Volgushev, A.N., N.V. Etkin and M.V. Paturoev. 1986. Svoistva, tekhnologiya i oblasti primeneniya polimercernykh betonov [Properties, technology and application fields of polymer sulphur concretes]. In: *Tsvetnaya Metallurgiya*. TsNIItsvetmet, Moscow, No. 8, pp. 53–55.

81. Volgushev, A.N. and M.V. Paturoev. 1985. Stoikost' sernykh betonov k vneshnim vozdeistviyam [Stability of Sulphur Concretes against External Influences]. Sb. dokl. Ashkhabad.

82. Vorob'ev, V.A. *et al.* 1978. Goryuchest' polimernykh stroitel'nykh materialov [Combustibility of Polymer Structural Materials]. Stroiizdat, Moscow, 225 pp.

83. Voznesenskii, V.A. *et al.* 1983. Sovremennye metody optimizatsii kompozitsionnykh materialov [Modern Methods of Optimising Compound Materials]. Budivel'nik, Kiev, 144 pp.

84. Zhavrid, S.S., V.I. Malikhtarovich and V.M Abramov. 1982. Stroitel'nye izdeliya i konstruktsii iz polimerfosfogipsa [Building Materials and Structures made of Polymer-Phosphogypsum]. Urodzhai, Minsk, 168 pp.

85. Zhurkov, S.N. 1967. Kineticheskaya kontseptsiya prochnosti tverdykh tel [Kinetic concepts of the strength of solid bodies]. *Izv. AN SSSR, Neorganicheskie Materialy*, **3**, pp. 1767–1775.

86. Zhurkov, S.N. and S.A. Abasov. 1962. Vysokomolekulyarnye Soedineniya [High-molecular Compounds]. **3**, 1961; **4**, 1962.

87. Zubov, P.I. *et al.* 1968. Issledovanie mekhanizma strukturoobrazovaniya napolnennykh poliefirov [A study of the mechanism of structure formation in filled polyesters]. *Kolloidnyi Zhurnal*, **30**, No. 3, 375–378.

88. Zubov, P.I. and L.A. Sukhareva. 1982. Struktura i svoistva polimernykh pokrytii [Structure and Properties of Polymer Coatings]. Khimiya, Moscow, 255 pp.

89. Zubov, P.I., L.P. Sukhareva and V.V. Paturoev. 1964. *Kolloidnyi Zhurnal*, **26**, No. 4, pp. 454–457.

Printed in India